T0313739

Wireless Coexistence

Wireless Coexistence

Standards, Challenges, and Intelligent Solutions

Daniel Chew
Johns Hopkins University Applied Physics Lab
Darnestown, MD, USA

Andrew L. Adams
Johns Hopkins University Applied Physics Lab
Darnestown, MD, USA

Jason Uher
Johns Hopkins University Applied Physics Lab
Baltimore, MD, USA

IEEE PRESS
WILEY

Published by John Wiley & Sons, Inc., Hoboken, New Jersey.
Published simultaneously in Canada.

For general information on our other products and services or for technical support, please contact our Customer Care Department within the United States at (800) 762-2974, outside the United States at (317) 572-3993 or fax (317) 572-4002.

Wiley also publishes its books in a variety of electronic formats. Some content that appears in print may not be available in electronic formats. For more information about Wiley products, visit our web site at www.wiley.com.

Library of Congress Cataloging-in-Publication Data:

Names: Chew, Daniel (Electrical engineer), author. | Adams, Andrew L.,
   author. | Uher, Jason, author.
Title: Wireless coexistence : standards, challenges, and intelligent
   solutions / Daniel Chew, Andrew L. Adams, Jason Uher.
Description: Hoboken, New Jersey : Wiley, [2021] | Includes bibliographical
   references and index.
Identifiers: LCCN 2021003859 (print) | LCCN 2021003860 (ebook) | ISBN
   9781119584186 (hardback) | ISBN 9781119584223 (adobe pdf) | ISBN
   9781119584124 (epub)
Subjects: LCSH: Wireless communication systems.
Classification: LCC TK5103.2 .C45375 2021 (print) | LCC TK5103.2 (ebook)
   | DDC 384.54/524–dc23
LC record available at https://lccn.loc.gov/2021003859
LC ebook record available at https://lccn.loc.gov/2021003860

Cover Design: Wiley
Cover Image: © Elnur/Shutterstock

Set in 9.5/12.5pt STIXTwoText by Straive, Pondicherry, India

MIX
Paper from
responsible sources
FSC
www.fsc.org   FSC® C013604

10   9   8   7   6   5   4   3   2   1

Contents

Author Biographies

Daniel Chew is a member of the Senior Professional Staff at The Johns Hopkins University Applied Physics Laboratory and teaches in the Engineering for Professionals program at Johns Hopkins University. He received a Bachelor's in Electrical Engineering from the University of Delaware, in 1998, and a Master's of Science in Electrical and Computer Engineering from Johns Hopkins University, in 2008. He has been a Lecturer at Johns Hopkins University since 2011. He is the author of the book The Wireless Internet of Things: A Guide to the Lower Layers (Wiley-IEEE, 2018). His professional interests include Radio-Frequency Machine Learning, Digital Signal Processing, the Internet of Things, and Software-Defined Radios.

Andrew Adams holds a B.S.E.E. from the University of Maryland at College Park, and an M.S.E.E. from Johns Hopkins University. He is currently a member of the Senior Professional Staff at the Johns Hopkins University Applied Physics Laboratory, and is a U.S. Navy veteran. Andrew also teaches Software Defined Radio at Johns Hopkins University. His professional experience includes Machine Learning for Signal Processing, Wireless Communications, Digital Signal Processing, and Radio-Frequency (RF) Hardware Design. His research interests include Intelligent Wireless Systems and Neuromorphic Computing

Jason Uher received his PhD from the University of Nebraska, in 2012, joining the Senior Professional Staff at The Johns Hopkins University Applied Physics Laboratory thereafter. Since then, has applied cutting edge research to the nation's critical problems across multiple networking domains. His research includes topics such as PHY/MAC layer security and anonymity, software defined radio processing techniques, system complexity analysis, and wireless networking protocol evaluation.

Preface

Book Motivation

Wireless communications has evolved considerably since we began practicing in the mid 1990s. Many of us started on Digital Signal Processors from Analog Devices, Texas Instruments, or Motorola, where we implemented basic signal processing functions in C using fixed-point math. If real-time processing requirements were ultimately not met, we downgraded specific functions to assembly language as needed. At the time, those were considered cutting-edge devices and standard implementation practices. Coming out of college, even with a Master's in Electrical Engineering, we knew just enough to get started, and had to learn "tricks of the trade" and "best practices," on the job, from more experienced Communications engineers. We are sure many have a similar background.

Fast forward to today, and the landscape is very different. The widespread availability of Software Defined Radio (SDR) hardware and software development environments have allowed students, early career engineers, practitioners in other disciplines, and maybe more importantly entrepreneurs to implement and deploy wireless capabilities much, much faster than was possible 20 years ago. Furthermore, advanced high-level tool chains allow us to implement signal processing functionality on Field Programmable Gate Arrays (FPGAs) and Graphical Processing Units (GPUs), with limited experience on either platform; something unheard of even 10 years ago. More recently, GPUs have been targeted for Machine Learning (ML) and other highly parallelizable computing operations. ML is increasingly playing a larger role in wireless communications applications, which we investigate in this text.

This lower threshold to fielding wireless capability has been in direct response to the explosion in demand for wireless services and the perceived lack of spectrum available to support these services. History has shown that regulatory bodies cannot solve this problem with static spectrum allocations; hence our need to devise ways to coexist with one another in the spectrum.

This certainly does not suggest or promote a wild-west scenario where spectrum allocations are routinely hijacked, but does suggest that where feasible, multiple users can coexist to maximize spectrum utilization without sacrificing quality of service. This is one of the tenets of Cognitive Radio, which is still a highly active research area today.

Many wireless standards have been published to help us coexist in the spectrum. The Third Generation Partnership Project (3GPP) and Institute of Electrical and Electronics Engineers (IEEE) are prime examples of standards bodies with extensive portfolios of published works. These represent thousands of hours of research to arrive at optimal solutions for targeted applications. There are too many to list here, but we do discuss select standards in this text. Why? We believe there is an intersection of domain knowledge, standards information, and practical experience, which helps engineers of all disciplines navigate a crowded spectrum. We acknowledge that much of this information exists in disparate sources, including many of the seminal texts relied upon in our own backgrounds, which are listed in the references of each chapter. However, we believe this text serves to pull many of these important concepts together as a toolbox of sorts. To this, we add our own voice, including practical experience in dealing with the many facets of wireless coexistence.

Audience

This text targets a wide audience, including both undergraduate and graduate electrical engineering and computer science students, practicing engineers in these disciplines, and practicing engineers in other disciplines relying on wireless communications to support their own applications. This text supports this diverse audience by introducing basic concepts in early chapters, and then building upon these in the later chapters focused on specific algorithms and wireless standards.

Book Organization

This text is organized as follows. Chapters 1–4 provide an introduction to various concepts fundamental to wireless coexistence. These chapters are especially helpful for those readers without a strong background in wireless communications. Chapter 1 introduces wireless coexistence, and serves as a foundation for ensuing chapters. Chapter 2 discusses concepts related to spectrum regulation and motivates the need for a new, flexible usage model. Chapter 3 provides an overview of the communications concepts referred to in the coexistence standards discussed in this text.

Chapter 4 expands upon this by introducing the multiuser concepts central to many wireless protocols.

Chapters 5 and 6 present advanced material related to secondary spectrum use. Chapter 5 introduces secondary spectrum use concepts and presents a mathematical framework for signal detection. Chapter 5 also derives several signal detection algorithms widely used in the standards discussed in this text. Chapter 6 discusses intelligent radio concepts and their application to wireless coexistence challenges. Chapter 6 also derives several algorithms fundamental to ML applied to wireless communications.

Chapters 7–9 provide an overview of the wireless coexistence mechanisms outlined in several prominent wireless standards. Chapter 7 introduces the IEEE 1900 series and explains its relevance to wireless coexistence. Chapter 8 does the same for the IEEE 802 series; maybe no standard series is more widely known or deployed. Chapter 9 introduces the wireless coexistence mechanisms defined in 3GPP Long-Term Evolution (LTE) License-Assisted Access (LAA). This is the first standard aimed at utilizing unlicensed spectrum for licensed carrier traffic.

We conclude with some forward-leaning concepts and final thoughts in Chapter 10.

Acknowledgments

The authors acknowledge the people who helped make this book a reality. The authors first thank the series editors Jack Burbank and Bill Kasch for giving us the opportunity to write this book. The authors also thank Joe Bruno, Miller Wilt, and Jared Everett for their extensive reviews on early revisions.

Dan dedicates this book to his wife, Lleona, and to his children, Marin, Everett, and Theodore. Their love and support made writing this book possible.

Andrew dedicates this book to his wife Michelle, and son Spencer. Your love, support, and patience make all life's endeavors possible. Thank you.

Jason dedicates this book to his wife Mary and "Captain" Larry for their unending support in all endeavors.

1

Introduction

It is common for both the general public and sophisticated engineers alike to take the concept of wireless communications for granted. Since the early research into the properties of the electromagnetic spectrum, scientists have sought understanding of the so called "Luminiferous Aether" [1]. Even with the vast amount of wireless devices in play today, the subject is often treated as a form of black magic: you put energy into a medium and it simply shows up where you want it. Though human understanding of the electromagnetic world has come a long way since the days of the Aether, there is still much that we do not understand about the propagation of electromagnetic waves in complex environments such as the natural and manmade landscapes we expect our wireless devices to operate in. As our understanding of the electromagnetic spectrum has changed over time, so have the methods by which we use that spectrum. While communicating over long distances was the original, and still most popular, use of spectrum there are now several aspects of an average person's day-to-day life that are made better by use of the spectrum. Outside of the communications role, we also use spectrum every day for things like sensing our environment, transferring energy from one place to another, heating objects, and many more. Every single one of these applications requires that the operator "use up" some amount of electromagnetic spectrum while accomplishing their goal. In order to understand why the efficient use of this spectrum is important, it is essential that we first understand what makes the electromagnetic spectrum a shared resource.

Wireless Coexistence: Standards, Challenges, and Intelligent Solutions, First Edition.
Daniel Chew, Andrew L. Adams, and Jason Uher.
© 2021 The Institute of Electrical and Electronics Engineers, Inc.
Published 2021 by John Wiley & Sons, Inc.

1.1 A Primer on Wireless Coexistence: The Electromagnetic Spectrum as a Shared Resource

In this section, we will seek to establish a fundamental baseline of understanding around wireless communications. This section is targeted at the wireless communications novices utilizing this book as a crash course in wireless coexistence standards. However, even seasoned RF scientists and engineers may find it useful as a number of key principles critical to the analyses in the remainder of the book are spelled out explicitly. This section, then, establishes a baseline of thinking from fundamental principles that can be used to reason about the how and why of coexistence from a consistent perspective. This common baseline allows for a like-to-like comparison when dealing with different styles of coexistence and gives the reader a consistent rubric for considering the potential tradeoffs of those styles.

1.1.1 Basic Description of Spectrum Use and Interference

When examining spectrum use, there are three orthogonal bases that are used to separate different users: time, space, and wavelength. When trying to understand these bases it is common to use sound as an analogous system to help reason about the properties of waves. Electromagnetic waves behave in many of the same ways that sound waves do, with the primary difference being which physical medium is excited with energy.

The first basis, time, is the easiest to understand. If someone is currently transmitting radio waves, they will be occupying the same portion of the spectrum at that time. The second basis, location, is similar but with a caveat. When someone transmits from a particular place, they are occupying the spectrum around that place. However, there are a number of factors that influence the degree to which they are *using* the spectrum there which will be discussed later in the propagation section. The third basis is the wavelength, or frequency, of waves used to transmit your radio signal. The signal sent through the air will always occupy a contiguous band of the spectrum; the width depends upon the amount of information being sent over the air. Figure 1.1 shows an example of a time-frequency map that might depict the transmitters in use at a given location. The larger a block is along the frequency axis, the more bandwidth it consumes at that time.

These types of maps can be helpful to visualize the spectrum usage by different users in a particular area. Which users are able to transmit in each section of a frequency band are typically displayed using a band plan.

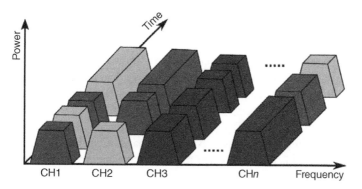

Figure 1.1 An example of a time frequency map.

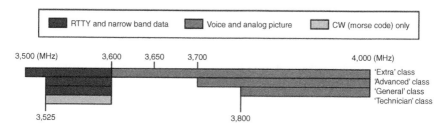

Figure 1.2 An example of a band plan.

For example, Figure 1.2 shows one of the United States amateur radio band plans for the 80 m allocations. Different users, separated into different classes by their capabilities, are allowed to use different sections of the spectrum according to this plan at any time.

One deficiency in a band plan is that it does not specify the physical location from which a user may transmit. This is typically enforced through a combination of the licensing authority and limits on the transmit power that a user can broadcast with. For example, broadcast AM radio stations transmit from known locations (where antenna the tower is), and are assigned a maximum amount of power they can use to transmit. Because the propagation characteristics of the AM Broadcast band are well understood, limiting the power to a certain level performs the same function as ensuring the signal will only be received within a given geographic area. Similarly, mobile users usually have the same power restrictions but have the additional restriction of ensuring that they are operating within a specific boundary, typically the jurisdiction of the licensing authority.

With these three potential bases, it is relatively easy to answer the question "*what is interference?*" IEEE 1900.1 [2] defined **Interference** as:

> In a communication system, interference is the extraneous power entering or induced in a channel from natural or man-made sources that might interfere with reception of desired signals or the disturbance caused by the undesired power.

Interference, in the context of wireless coexistence, means impairing the transmission of another user. This is caused when multiple users operate at the same time, within the same bandwidth, and in the same geographic location as each other, with no means to de-conflict that resource. Chapter 4 will discuss in depth several multiple access strategies such as Code Division Multiple Access (CDMA), which is intended to allow concurrent use of a spectrum resource in time, frequency, and space; but suffers from *interference* caused by the multiple independent users on one spectrum resource. This is type of interference called Multiple User Interference (MUI). Chapter 10 will expand on that discussion and delve into Non-Orthogonal Multiple Access (NOMA), which revolves around concurrent use of spectrum resources and the interference this causes.

1.1.2 Understanding What It Means to Occupy a Band

Using the earlier definition of interference, "operating at the same time, within the same frequency band, and in the same geographic location as another user," leaves a number of practical questions about what it actually *means* to be in the same band or the same place. The logical representation of band usage, such as that shown in Figure 1.1, shows an idealized representation of what it means to occupy a band. In reality, wireless transmissions are a physical process that do not cleanly start and end at the exact edges of the allocated band. A more realistic interpretation of the spectrum usage is shown in Figure 1.3.

Due to the physical nature of modulating data onto signals, it is impossible for a transmitter to keep all of the energy only in the band of interest. This means that there will always be interference outside of the allowed band, even if it is only a very small amount.

1.1.3 Spectral Masks

It is an unfortunate fact of radio communications that an information-carrying signal cannot be made to occupy a finite bandwidth. Because transmitted signals will always produce some amount of noise outside the primary transmission band, the majority of the regulations and licensing requirements in place today focus on ensuring that the interference introduced outside the allocated bands is limited to a manageable amount. These limits are usually defined through the use of what are called **Spectral Masks**. A spectral mask outlines the amount of power that a licensee's devices can

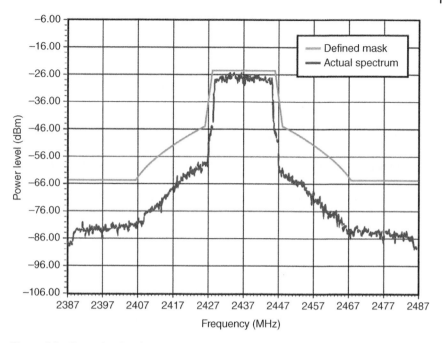

Figure 1.3 Example of realistic spectrum usage for one signal.

radiate over bandwidth. Power over bandwidth is called the **Power Spectral Density** (PSD). Spectral masks are determined from a variety of constraints including but not limited to international regulations, other users in the same band or adjacent bands, and the likelihood of interference with other equipment. For example, Figure 1.4 shows the spectral mask imposed on 802.11 devices when using Direct-Sequence Spread Spectrum (DSSS) [3].

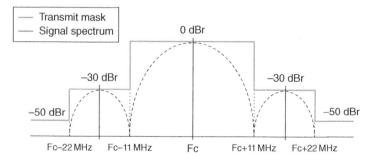

Figure 1.4 Example of a transmit mask. *Source*: IEEE 802 [3].

The bold line in Figure 1.4 is the spectral mask and it limits the power spectral density of the signal that may be transmitted in terms relative to the peak as a function of frequency. The limit is imposed in decibels relative to the reference level (dBr). 0 dBr is at the *reference level*. The reference level is the highest spectral density of the transmission. The channel, in this context defined as the intended transmission bandwidth, spans from ±11 MHz of the center frequency. Outside the intended transmission range, ±11 MHz from the channel center frequency, the power spectral density must not exceed −30 dB relative to the peak power spectral density. Past ±22 MHz from the center frequency, the radios are only authorized to transmit below −50 dBr. By imposing this limit past ±22 MHz from the center frequency, the spectral mask attempts to limit interference caused to other wireless systems. This is a typical representation of a spectral mask, showing the allowed interference that a transmitter can introduce to the signal around it. However, in some cases, masks can be much more complicated than simply *in-band* and *out-of-band* emissions. For example, Figure 1.5 shows the spectral mask for the FCC's amendment to the Part 15 code that allows use of "ultra wide band" unlicensed devices.

This mask shows that devices compliant with this regulation will not have an overall radiated PSD of −41.3 dBm/MHz across the majority of the band with two main exceptions. First, transmitters must not radiate more than −75.3 dBm/MHz in the band used by the Global Positioning System (GPS), which is highly sensitive to local noise from other ground-based devices. Second, these devices are limited to −61.3 dBm/MHz in bands where other sensitive satellite communications networks and radio astronomy receivers typically reside.

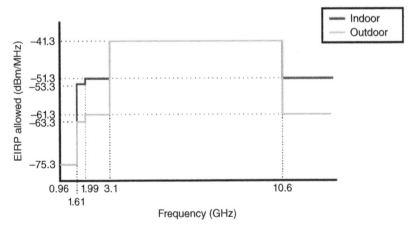

Figure 1.5 US FCC Part 15 spectrum mask. *Source:* Based on Radio Frequency Devices [4].

In both cases, the UWB transmitters would overpower the relatively weak signals coming from space and therefore these critical services are protected from UWB devices with a stricter mask in the bands that they typically operate.

1.1.4 Bandwidth and Information Rate

While band plans and spectral masks might tell a user how much spectrum they are able to occupy, what does that actually mean for the user in real world terms? Practically speaking, the more spectrum or bandwidth that a user consumes, the more information they are able to convey. Information, in this case, can be anything that conveys meaning from one user to another, whether that be analog voice communication, a video signal, or digital data of any type [5]. The amount of information that can be sent within a specific chunk of spectrum is limited in theory by the Shannon Limit [5], which says that there is an upper limit on the amount of information that can be conveyed through any given bandwidth (*channel*). Ultimately, the conclusion is that as more spectrum is assigned to a given user, that user will be able to send more information.

1.1.5 Benefits of Different Frequencies

One of the major concerns different users of the wireless spectrum have revolves around the frequencies of electromagnetic waves they are allowed to utilize. The utility of spectrum for various communications and sensing applications is not just about *how much* but also *at what frequency*. For example, if a user is granted 10 MHz worth of spectrum at a very low frequency such as 30 kHz, the way that spectrum can be used is drastically different than how 10 MHz of spectrum at a shorter wavelength in the 300 MHz range can be used. One major factors that affects all communication systems is antenna size. In order to be efficient, the size of the antennas in a radio system are directly impacted by the frequency used. For example, operating at a very low frequency requires a large antenna that would make it unsuitable for mobile operations. In addition, different propagation characteristics might make one band more useful than another for different applications. For example, some frequencies can travel extremely far around the globe by bouncing off the Earth's ionosphere. While this makes that frequency range very useful for terrestrial communications, it would be nearly useless for satellite communications as the signal would not be able to escape the atmosphere. Similarly, frequencies that are absorbed by walls and buildings would not be good for city-wide communications but make excellent candidates for in-home wireless networks.

In addition to the propagation characteristics of different frequencies, the maximum data rate that can be used with practical circuitry goes up with the carrier frequency used [6]. This means that if a user wants to send a lot of data in a short amount of time, a radio can be more easily built if it utilizes a higher carrier frequency. In general, this makes the higher frequency bands more valuable as data rates and overall system bandwidths grow alongside user appetites for data.

1.2 The Role of Standardization in Wireless Coexistence

Wireless communication standards are an agreement between users of a shared wireless link to communicate in a predictable way that allows all users to both share the spectrum and understand one another. This agreement includes both physical concerns such as the modulation scheme, bands of operation, and data rate as well as the instructions for negotiation of channel use and network formation. Having such a common set of definitions for a given wireless link is a prerequisite for interoperability between devices designed and manufactured by different vendors. Therefore, wireless standards are necessary in order for wireless communications to be viable in a large marketplace with many end-users.

The average end-user may not know of the importance of these standards. What the end-user is concerned with is that the product they have purchased works, that it satisfies their needs. For wireless devices, the user will need their purchased device to connect with other devices and base stations in order to establish a wireless link. The interoperability of all these devices rests upon conformance to a wireless standard. It can therefore be said that without the wireless standards defining bounds for these devices, there is no wireless link and therefore no wireless connectivity. Ultimately, meeting the user's needs is what drives the standards.

IEEE 802.11 is an excellent example of this need for standardization. Since the inception of that standard in the late 1990s, Wireless Local Area Network (WLAN) connectivity has become ubiquitous. Without the IEEE 802.11 standard, Wi-Fi connectivity could not be seen as a common amenity offered by hospitality and retail services.

The above argument establishes the need for standardization in wireless communication links; but what about wireless coexistence? Section 1.1 introduced the concept of spectrum regulation. Why is not spectrum regulation, as dictated by a government authority, sufficient to provide coexistence? One example of the need for industry associations to define coexistence standards can be seen among wireless service providers such as cellphones. As will be seen in Chapter 4, which tackles mitigating contention across equal priority users, wireless service providers must share their allocated spectrum resources across many subscribers. There are many end-user devices attempting to access the same licensed spectrum resource. Therefore, these wireless service providers must define a common means of accessing those limited spectrum resources. Once again, it is the need of the application which drives the standardization.

Standards for wireless communication links are often organized into **protocol stacks**. A protocol stack is a series of layers of processing, each *stacked* upon the other. The functions necessary to establish and maintain the wireless communications link are divided across those layers. The seven-layer Open Systems Interconnection (OSI) model [7] provides a common ontology for the layers and their functional components. As will be seen in Chapter 4, standards for wireless communication links must often define a **Medium Access Layer** (MAC) that dictates the rules for devices wishing to access shared spectrum resources. However, relying on the MAC layer alone to resolve the problem of increasing contention in spectrum access will not be sufficient in the growing complexity of the wireless marketplace.

Reference [8] provides a survey of spectrum sharing, specifically across different technologies. Figure 1.6 illustrates a "technology circle" based on the seven-layer Open Systems Interconnection (OSI) stack. Layers 1 through 7 are the traditional seven layers of the OSI stack. Layer 0 represents government regulations. These government regulation (layer 0) drives the design decisions of a physical layer of a communications standard (layer 1).

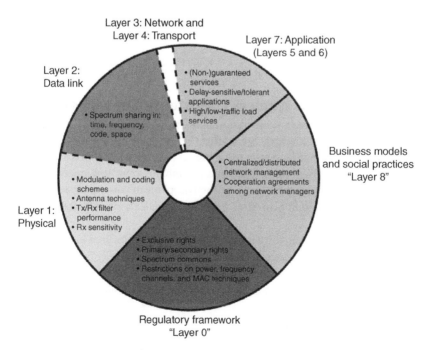

Figure 1.6 The technology circle of wireless coexistence. *Source:* Voicu et al. [8]. © [2020] IEEE.

Layer 8 represents standard practices of the users of wireless systems. It is the needs of the end-user (layer 8) that drive the regulations (layer 0). This connection between layer 8 and layer 0 closes the circle shown in Figure 1.6.

The key concept that the user should take away from Figure 1.6 is that the end-user applications are driving the need for more wireless coexistence. Wireless devices are already ubiquitous, and the volume of wireless devices in marketplace keeps growing. As the demand for more spectrum access grows, congestion across the spectrum grows more severe. As the problem of congestion in spectrum access grows more complex, the mechanisms providing coexistence grow more complex. An example of this is the wireless standard IEEE 802.15.2. This standard did not define a wireless communications link, but rather, it provided recommendations to alleviate contention and interference between two very popular wireless communication links.

Wireless coexistence strategies are a necessity to mitigate congestion in spectrum access. Over the past few decades, these strategies employed have grown more complex. The next section will detail those strategies, and provide a framework for grouping them. Wireless coexistence has developed from a strategy of simply keeping all emitters physically separated, to collaborative spectrum-sensing and spectrum re-use. These increasing complex wireless coexistence strategies require coordination and compliance among multiple devices. The best way to achieve that goal is to provide the vendors of those devices a wireless coexistence standard.

1.3   An Overview of Wireless Coexistence Strategies

There are many potential methods for sharing the same spectrum resources. This act of communicating simultaneously using the same spectrum resources is called "wireless coexistence." By working together and applying some fundamental communications principles, everyone can use the spectrum to communicate, sense, and transfer energy without worrying that their applications will fail due to interference.

This book will cover in depth the wireless coexistence strategies used in a variety of standards but different strategies can be categorized more generally by the method they use to deconflict the spectrum resources. The five primary strategies for deconfliction are Separation, Mitigation, Monitoring, Sensing, and Collaboration.

1.3.1   Separation Strategies

Multiple access schemes are an example of a separation strategy. Numerous multiple-access schemes exist. The ones in vogue have changed over time. For the first generation cellphones (1G), **Frequency Division Multiple Access** (FDMA) was used. As the density of cellphones per person increased, the second generation (2G) of cellphones employed **Time Division Multiple Access** (TDMA) to increase the number of users any one cellular base station could service. TDMA can transmit to multiple user-nodes on one carrier frequency, avoiding intermodulation distortion. The time-slotted channels could be more easily allocated than frequency-channels.

As the number of cellphones increased, **Code-Division Multiple Access** (CDMA) schemes were introduced in the third generation (3G) of cellphones. CDMA offered a new dimension (spreading codes) for channelization; and these spreading codes offered improved resilience to frequency-selective multipath and errant emitters (interferers). This "new dimension" meant that the individual nodes were interfering with one another. CDMA is not simply a channelization method in the traditional sense, but also a mitigation method.

The fourth generation (4G) of cellphones serviced even more users through **Orthogonal Frequency Division Multiple Access** (OFDMA) whereby small blocks of time and frequency resources on any one carrier could be directly assigned to a user.

One key feature these separation strategies share is that the system "owns" the spectrum in which it operates. There is an expectation of exclusivity, and a regulatory authority guarantees that exclusivity. Interference need not be mitigated because any interfering signal would be in violation of that regulatory authority.

These separation strategies will be discussed in more detail in both Chapters 2 and 4. In Chapter 2 the role of spectrum regulation will be addressed. In Chapter 4 separation will be explored as the primary existing means of alleviating contention between users of equal priority.

A separation strategy is insufficient in an unlicensed band like the Industrial, Scientific, and Medical (ISM) 2.4 GHz band. For successful operation in an unlicensed band, interference must be expected and some mitigation employed.

1.3.2 Mitigation Strategies

Mitigation strategies expect interference. These strategies include in the modulation scheme some mechanism to ameliorate the ill effects of interference. Examples of such a strategy include spread spectrum techniques. Two spread spectrum techniques seen in the 2.4 GHz ISM band are direct-sequence spread spectrum and frequency hopping. Spread spectrum techniques are the norm for devices operating in the 2.4 GHz ISM band. Spread Spectrum techniques will be discussed in Chapter 4.

1.3.3 Monitoring Strategies

When interference between signals causes problems in the wild, link monitoring and adaptive mitigation techniques can be used to share the spectrum. Frequency-hopping systems like Bluetooth employ "adaptive frequency-hopping" [9]. The gist of the concept of adaptive frequency-hopping is for the devices to automatically "blacklist" certain frequency channels that prove problematic. This requires the devices to *sense* which frequency channels either already have activity on them or which ones are rendered unusable by multipath. The difference between adaptive systems which require monitoring of the wireless link (monitoring strategies) and systems which simply march through any encountered interference (mitigation strategies) will be shown in both Chapters 4 and 8.

1.3.4 Sensing Strategies

Sensing strategies include concepts such as spectrum sensing and dynamic spectrum access where time-frequency resources that can be used are determined before transmission. This is distinct from the multiple access schemes discussed in the separation strategies because it is the individual nodes themselves performing the sensing operations. The individual nodes attempt to find **white space**, which is unused spectrum, within a target band of operation. This sensing may be augmented by accessing a database of known emitters in the geographic region of operation.

A common approach is simply to "listen before talk" which involves sensing for the presence of another active emitter in a frequency channel before transmitting. This approach is called **Carrier Sense Multiple Access** (CSMA). CSMA is employed in numerous standards including IEEE 802.11 more commonly known as Wi-Fi.

Examples of wireless standards defining systems which operate in white spaces are IEEE 802.22 [10] and IEEE 802.11af [11]. In both of those standards, spectrum sensing and rules for co-existence are defined in the individual standard. So long as a network of emitters adheres to one standard or the other, the emitters will be able to coexist and dynamically make the best use of the available spectrum.

1.3.5 Collaboration Strategies

A collaborative coexistence strategy is one that relies on different wireless systems sharing and exchanging information. Such collaboration can be seen in the standards IEEE 802.19.1 [12], IEEE 802.15.2 [13], and IEEE 802.22 [14]. In collaborative strategies, emitters share information between each other and coordinate spectrum use. If the emitters are coordinating access, it may not be necessary to employ any spectrum sensing or interference mitigation.

As will be seen in the standards covered in this book, collaborative strategies are often paired with other strategies. The results are that when one unit senses spectrum in one location, it can share that information with a larger network.

1.3.6 Combining the Strategies

The strategies discusses in this section need not be exclusive. A sensing strategy may also employ a mitigation strategy. A wireless system may have exclusive access to a specific bandwidth, and that exclusivity would be an example of a separation strategy, but within that system there may be licensed concurrent emitters employing strategies to coordinate and avoid interference.

1.4   Standards Covered in this Book

This book will provide detailed overviews of selected wireless coexistence standards and delve into the background theory behind those standards. The wireless coexistence standards to be covered will include:

- IEEE 1900
- IEEE 802.22
- IEEE 802.11af
- IEEE 802.15.2
- IEEE 802.19.1
- And LTE LAA

IEEE 1900 is a series of standards that have been developed and refined since 2005. This series of standards is currently maintained by Dynamic Spectrum Access Networks Standards Committee (DySPAN-SC). Among the series, one standard that will be referenced throughout this book is IEEE 1900.1 which provides a common set of terminology and definitions for wireless coexistence. The IEEE 1900 series is detailed in Chapter 7.

IEEE 802.22 and **IEEE 802.11af** are both wireless standards for communication links in the television band. As such, these standards define links for *secondary users*. The concept of tiers of users will first be addressed in Chapter 4. Spectrum sensing will be addressed in Chapter 5. Intelligent Radio functions will be addressed in Chapter 6. These two standards will be discussed as wireless coexistence standards in the IEEE 802 series in Chapter 8.

IEEE 802.15.2, as briefly discussed in Section 1.2, provides a recommendations for wireless coexistence between the Wireless Personal Area Network defined by IEEE 802.15.1 (now Bluetooth) and the IEEE 802.11 WLAN (Wi-Fi). IEEE 802.15.2 has been officially withdrawn, however, some of the recommendations made in that standard were adopted and are still present to this day. That makes IEEE 802.15.2 an important part of the history of the development of wireless coexistence standards. This standard will be discussed as wireless coexistence standards in the IEEE 802 series in Chapter 8.

IEEE 802.19.1 specifies a means of coexistence for nodes which do not otherwise share any standardization. This standard can be seen as providing an overlay onto the MAC layer of other standards. One example would be to mitigate contention between 802.22 and 802.11af which would both vie for secondary user access to TV Band white space.

UMTS Long-Term Evolution License Assisted Access (**LTE LAA**) is part of release 13 of the LTE standard. LTE LAA is a variant of *LTE-Unlicensed* in which LTE would operate in unlicensed bands. LTE is a ubiquitous telecommunication system, and it is operation in potentially congested unlicensed bands is relevant to the future of wireless coexistence. LTE LAA will be detailed in Chapter 9.

1.5   1900.1 as a Baseline Taxonomy

One of the biggest hurdles when trying to discuss wireless technologies in a coherent fashion relates to the idea of language. Technical terms very often mean different things to different people leading to mass confusion and generally unproductive discussion when it comes to comparing and contrasting different regulatory schemes. For example, what does it mean for a radio to be *Software Defined*? While most would agree this means that the radio functionality is programmed with some sort of software description, the gamut of potential interpretations is extremely wide. In a similar fashion, one could ask what it means to be a *Cognitive Radio*. Clearly, at its core, the term *Cognitive Radio* means just one thing: that the radio is using some intelligent decision-making processes to operate. Over time, however, the term cognitive radio has become muddied to include a radio that may or may not utilize things like dynamic spectrum access, spectrum monitoring, radio signal identification and more. While it certainly possible that a radio could use cognition to perform these tasks individually or, alternatively, use the results from these task as inputs to a cognitive process they are not necessarily, in and of themselves, what make a radio "cognitive." It is critical that a well-defined taxonomy is used when comparing wireless coexistence mechanisms in a fair and reliable fashion. A good taxonomy will allow users to separate the irrelevant policy and system architecture differences and focus efforts on comparisons that tease out the core benefits of different schemes.

Although the IEEE 1900 standards will be discussed in more depth in Chapter 7, a focused discussion of the 1900.1 standard here has the benefit of providing a general taxonomy for wireless coexistence technologies.

One of the critical contributions of the IEEE 1900 standards is providing a baseline taxonomy for terms and concepts relating to wireless coexistence. The definitions and concepts laid out in 1900.1 allow the IEEE to define exactly what does and does not constitute dynamic spectrum access but also provides a wider categorization of coexistence strategies in general. This idea is clearly presented from the IEEE 1900.1 scope statement:

> This standard provides definitions and explanations of key concepts in the fields of spectrum management, spectrum trading, cognitive radio, dynamic spectrum access, policy-based radio systems, software-defined radio, and related advanced radio system technologies.

Through a combination of the definitions themselves and the informative annexes, the reader can use this model to classify any spectrum access scheme. This includes both concepts in wide use today, such as fixed frequency allocations and the ISM unlicensed bands, as well as future concepts for spectrum access that are being proposed to reduce spectral inefficiencies. The formal definitions described in 1900.1 provide five categorical breakdowns that can be used when describing any radio system including device architectures, devices capabilities, command and control concepts, radio network types, and spectrum management concepts.

1.5.1 Advanced Radio System Concepts

The first category defined in 1900.1, "Advanced Radio System Concepts," provides a clean delineation between the terms surrounding radio systems. Figure 1.7 shows an Euler diagram of the advanced radio system concepts taken from the IEEE 1900.1 supplemental material which provides a useful visualization to the concepts discussed in the rest of this section.

The first primary distinction surrounds the concept of delineating which radio functions are implemented in either hardware or software. In order to provide a more accurate taxonomy, 1900.1 separates the idea of how radio functions are *defined* versus how they are *controlled*. For example, it is entirely possible for radio software to switch between different hardware signal processing functions as part of an adaptive spectrum access scheme. Separating the taxonomy alleviates some confusion about whether or not a system is considered a *software-defined radio* from a regulatory point of view if the radio functions are implemented in hardware but dynamic control occurs in software. In the IEEE 1900 taxonomy, the terms *Hardware Radio* and *Software Defined Radio* (SDR) imply that the signal processing functions themselves are implemented in either hardware or software respectively and whether they can be changed without physically modifying the device. If the signal-processing functions are implemented in software and "cannot

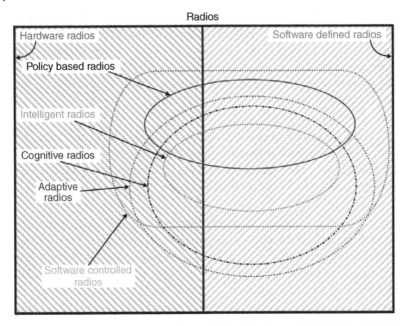

Figure 1.7 Radio classification guide for IEEE 1900. *Source*: I. S. 1900.1-2019 [2].

be changed in the field, post manufacture, without physically modifying the device" then the device is still considered a hardware radio.

Additionally, a radio can be a *Software Controlled Radio* (SCR). In this case, in addition to the signal processing functions themselves, the idea of "control" is separated from the radio and "implies more dynamic operational flexibility from radio interfaces, in contrast with software-defined radio which might in some cases be seen as a static implementation of the radio interface by software."

The second set of definitions outlined in advanced system concepts relates to how the radios behave. In the 1900.1 taxonomy, radios whose functions can be changed by software to support changes in their environment are called *Reconfigurable Radios*. These radios can be *adaptive, cognitive, intelligent, policy based*, or a combination of all four depending on how they are reconfigured. For example, an *Adaptive Radio* changes its operating parameters in response to changes in its environment. The key point is that how it makes those changes does not necessarily have to be done in an intelligent way. For example, changing operation based on time of day or set thresholds on bit error rate can be adaptive actions, but they are simple reactions and not made by reasoning about the environment. This is in contrast to a *Cognitive Radio* which "exploits cognition to control its behavior." If a cognitive radio is using machine learning in addition to simple cognition, it can be referred to as an *Intelligent Radio*. The final behavioral control type is referred to as a *Policy Based Radio*. In this case the radio's behavior is dictated by a "policy," a set rules and regulations that the radio manufacturer agrees to abide by. Realistically, all current radios are *policy based radios* in the sense that they are typically only allowed to transmit in certain bands a set a policy limit. In this case, such as with a hardware radio, the policy in enforced by the fact that the radio is incapable of transmitting outside of the policy rules and remains in effect for the lifetime of the device. For future dynamic spectrum access radios, however, this policy may be enforced for only a short time or only in specific locations. IEEE 1900.1 makes this distinction clear in the supplemental material by stating that the policy definition may occur during manufacture of the device, during deployment, or via real-time control.

1.5.2 Radio Capabilities

The second set of definitions provided in the 1900.1 standard are the different capabilities that a radio may have when accessing spectrum. The 16 individual capabilities defined in the standard can be loosely grouped into three categories: system actions, system inputs, and cognitive abilities. System actions are things that the radio can do. This category includes all actions that the radio can take when adapting to the environment. It includes the ability to adaptively change the transmission parameters of the radio such as transmit power and modulation in response to external stimuli such as bit rate or link quality. It also includes the ability to be agile in frequency, changing where in the spectrum the radio is operating on demand. System inputs are attributes about the system that can be used by higher-level control software to access the spectrum. These include the radio's ability to maintain a representation of its own state and the environment around it. This may include keeping track of other spectrum users in range, the location of the radio both in terms of physical space and regulatory authority, and other environmental attributes that the radio can use to reason about what policies can be executed. The last group of attributes are the cognitive abilities of the system. These include the ability to reason about how and when to change the operation of the radio in order to access the spectrum according to the allowed policies. This may include learning about the RF environment as it changes over time, consulting regulatory policies, and reasoning about the correct course of action in a given situation.

1.5.3 Network Types

IEEE 1900.1 differentiates three types of advanced radio networks depending on the capabilities of the network as a whole to operate in a given environment. The first, a **Cognitive Radio Network** is a network where the behavior of the radios is adapted based on a cognitive process running in the network. It is noted that the individual radios in this type of network do not necessarily have to be cognitive themselves, just that they are controlled by a cognitive process. For example, a central cognitive node could collect the environmental awareness information from the individual nodes and instruct the network to adapt. The second network type is **Dynamic Spectrum Access** networks. These are networks that adapt their spectrum utilization in real time according to changes in the environment and predefined policies. The final network type, **Reconfigurable Networks**, is any network that can be configured on the fly. It is important to note that these networks do not necessarily need to be made up of adaptive or reconfigurable radios themselves. For example, in a network with fixed configuration base stations, the network could be reconfigured at a higher layer to route traffic over the best RF link available at the time without actual changes made to radios themselves.

1.5.4 Spectrum Management

Of all the terms and definitions set forth in IEEE 1900.1, those defined in Section 7 relating to spectrum management bring the most clarity to this confusing topic. Clear and accurate descriptions of what it means for a radio to have a *frequency allocation* or what the difference between radio networks that *collaborate* versus those that *cooperate* provide a common baseline for discussing different protocol aspects in reasoned way. This section will cover the most useful terms and definitions outlined in the "Spectrum Management Definitions" section of 1900.1 with the intent that these terms will be used coherently for the rest of the book when discussing particular technical aspects of different protocols or different wireless coexistence schemes.

1.5.4.1 Basic Terminology

Section 7 of 1900.1, "Spectrum Management Definitions," outlines the basic terminology that frames who the actors are and what their goals are when it comes to wireless coexistence. In fact, 1900.1 even provides a formal definition for coexistence which is defined as "the state of two or more radio devices or networks existing at the same time and at the same place in a shared spectrum space."

1900.1 establishes that a **Coexistence Mechanism** is any technique that different radios might use to avoid causing interference with one another. Interference includes any energy, both natural and man-made, in the operating bandwidth of a user. Mitigating harmful interference is the primary focus of Spectrum Management. Through effective spectrum management, the Spectrum Owner can establish a set of rules and procedures, known as **Spectrum Etiquette**, that meet the spectrum owner's goals for that section of the spectrum. This etiquette could be very simple, such as in traditional licensing schemes dictating that only one license holder is allowed to use the band at all, or very complicated requiring users to adhere to any number of policy restrictions in order to access the spectrum. In addition, different etiquettes can be applied at different levels of the spectrum management hierarchy. For example, the ITU might define an allocation that specifies a high-level purpose for a given band which is then further restricted by federal or regional regulators. If the spectrum etiquette is adhered to, the final outcome should be the state of Electromagnetic Compatibility (EMC), which is when radios utilizing in the same TLF space can operate "without causing or suffering intolerable or unacceptable degradation."

1.5.4.2 Spectrum Access, Sharing, and Utilization

Ultimately, the goal of wireless coexistence is to ensure that there is fair and equitable access to a finite set of spectrum resources without interfering with one another. Because these resources are limited, how different users access and share the spectrum can dramatically affect the ability of users to meet the goals of the spectrum owner. In this paragraph we will outline the definitions from 1900.1 that provide what it means to use spectrum resources.

The first part of using spectrum resources is the concept of access. Spectrum Access is defined in 1900.1 as "Transmission or reception on the radio spectrum." The standard makes it clear that even though transmission in a band is the outwardly obvious "use" of the band, preventing harmful interference to licensed receivers is just as important. The distinction is important because wireless coexistence mechanisms often rely on the fact that if there is no transmitting entity in the band that the band constitutes a *free* chunk of spectrum. This is not always the case and the standard makes it clear that protected receivers are "accessing the spectrum" just as equally as a transmitter would. When users are actively accessing the spectrum, either while transmitting or receiving, that section of spectrum is being **utilized**. Note that there is no distinction between whether the user is authorized to transmit there, the noise is from a natural source, or if the transmission parameters do not follow the etiquette laid out by the spectrum owner: the spectrum is "utilized" regardless. 1900.1 defines the utilized spectrum, expressed in Eq. (1.1) as U, as the product of the occupied bandwidth (B), the physical space being utilized (S) and the time period (T) that the access takes places over.

$$U\left(\text{Hz} \cdot \text{m}^3 \cdot \text{s}\right) = B(\text{Hz}) \cdot S\left(\text{m}^3\right) \cdot T(\text{s}) \tag{1.1}$$

From this definition, we can see that any TLF space that is not currently utilized is a *free resource* that is available for another user. In 1900.1 this free resource is referred to as Spectral Opportunity and is defined as any frequency segment that "satisfies availability criterion" for access based on the spectrum etiquette in play at that given TLF block.

The concept of utilization defined above leads directly into the idea of ensuring that the spectrum is being used to its maximum benefits, known as the Spectrum Efficiency. One of the primary metrics for determining if spectrum is being used efficiently is the Spectrum Utilization Efficiency (SUE) metric, expressed in Eq. (1.2), which is defined as the amount of information transferred (M) to the amount of spectrum that is used by the spectrum access (U).

$$\text{SUE} = \frac{M}{B \cdot S \cdot T} \tag{1.2}$$

While the utilization of the resources is easily defined, the "amount of information transferred" is not necessarily clear cut. For this reason, the value of M should be adjusted to the value that makes sense for system design such as "raw data rate" (bits/Hz), Erlangs, "Number of radar channels available," etc. The goal of any future spectrum etiquette rules, then, should be to ensure that the spectrum owner is working to maximize the spectral efficiency of their bands.

One way to ensure that spectral efficiency is maximized is to implement an etiquette that allows multiple users to coexist in a shared spectrum space, defined as **Spectrum Sharing**. Technically speaking, both current models of frequency allocation are "spectrum sharing etiquette" albeit, highly simple ones. For example, in the unlicensed bands, the ISM style of granting anyone equal rights to frequency band is known as Horizontal Spectrum Sharing. Here everyone is allowed to access the band without regard for other users. In contrast, the traditional Spectrum Leasing model in which a single user has the right to access is known as Vertical Spectrum Sharing. Under vertical sharing, different users have different rights when it comes to accessing the spectrum. In this case the band is "shared" in the sense that one user has unfettered rights to access the spectrum and all others have no rights to access. The traditional model also exemplifies the concept of Frequency Sharing, a sharing paradigm in which set chunks of spectrum are allocated to specific users.

1.5.4.3 Dynamic Access and Spectrum Management Strategies to Improve Spectral Efficiency

While the current model of fixed frequency leasing is easy, it is certainly not efficient. It is clear that in order to improve spectral efficiency as a whole the traditional concepts surrounding fixed access will have to be amended and improved upon. **Dynamic Spectrum Access** is defined as "The real-time adjustment of spectrum utilization in response to changing circumstances and objectives." However, even though a radio may be capable of dynamic access, the spectrum is still managed in such a way that it may not be allowed to dynamically access certain resources. The set of rules that radio uses when deciding when and where to access the spectrum are referred to as a "policy" which is defined by the Policy Authority, the entity that has jurisdiction over spectrum usage. The ability to ensure that spectrum is properly managed via these policies called Policy Traceability, which says that all actions taken by dynamic spectrum access radios have evidence to support how that action meets the spectrum management policies. Through a combination of well written policies and clear etiquette on cooperation dynamic access to the spectrum will improve overall efficiency without generating harmful interference to existing technologies.

1.5.4.4 Collaboration versus Cooperation

By allowing radios to respond to the changes and react accordingly, spectrum opportunities can be exploited to provide an overall increase in spectrum efficiency without harmful interference to other users. From the perspective of 1900.1, these dynamic changes can be made through two subtly different methods: cooperation and collaboration. The first method, **Cooperation**, ensures that all the tasks necessary for co-existence are carried out by the entities engaging in dynamic spectrum access. This cooperation does not imply that the nodes are in communication with one another. For example, 802.11h mandates that 802.11 devices sense their environment before using a candidate channel for wireless networking [15, 16]. In this case the nodes are cooperating, because the 802.11 devices will not interfere with the primary users of the band but no formal communication occurred between the band users. In contrast to cooperation, nodes that are **collaborating** will actively communicate with one another, sharing spectrum sensing data and agreeing on the best deconfliction strategy. In general, collaboration would be preferred to cooperation because it allows the nodes to properly exchange their intentions and requirements in order to make an optimal plan for spectral use. However, in the case of legacy incumbent users who may not be capable of collaborating, cooperation allows newer radios to avoid the legacy users and still improve the spectral efficiency overall.

1.5.4.5 Fixed versus Dynamic Management

Obtaining access to the spectrum is dependent on the rules and regulations that are in force for any given TLF resource. For dynamic spectrum access, 1900.1 differentiates between access that is enabled via traditional spectrum management strategies and those that are enabled by newer, currently unimplemented, dynamic management concepts.

For traditional fixed allocation management, there are a number of methods by which dynamic spectrum access can be enabled. The first method is to allow Hierarchical Spectrum Access in which "...a hierarchy of radio users or radio applications determines which radios have precedence." This concept reflects traditional spectrum sharing research model that includes a *primary user* and *secondary user* where the primary user has full access to the spectrum and the secondary user may access the channel provided that they will not interfere with the primary. This type of access is split into two different categories: overlay and underlay. **Spectrum Overlay** describes a system in which the secondary user monitors for and exploits spectrum opportunities through Opportunistic Spectrum Access. In this way, the secondary user fits "between" the unused resources left by the primary user, and manages to operate without harmful interference. In contrast, **Spectrum Underlay** allows secondary users to transmit at any time but restricts the amount of interference generated to some preset threshold that occurs below the level of harmful interference to the primary user.

Another method of fixed management relies on cooperation between different spectrum owners. **Spectrum Pooling** is defined as the situation where multiple spectrum owners pool their resources to ensure more efficient use of the spectrum overall. For example, if mobile telephony operators in geographically distinct regions each held licenses for fragments of band, they could agree to pool those resources and utilize this newly formed contiguous chunk of spectrum without interfering with one another. Finally, the concept of **White Space Spectrum Band** builds on the hierarchical and spectrum overlay concepts by providing a method for accessing unused spectrum via formal means. When White Space, a spectrum allocation that is not currently being used by the owner, is available it can be added to a list of unused resources called a White Space Database. By compiling the geographic information in the database, secondary users can determine which spectrum resources are available for use in their area. This style of management often requires that secondary users utilize spectrum sensing to detect potential incumbents and regularly update their database information to minimize the risk to primary users.

In the future, spectral efficiency could see large gains through the use of a more flexible management style. **Dynamic Spectrum Management** is "A system of spectrum management that dynamically adapts the use and access of spectrum in response to changing circumstances and objectives." In this case, some method of collaboration between the users ensures that the spectrum is being utilized as efficiently as possible, adjusting the spectrum management policies as necessary to meet that goal. In order to adjust for these changes in the policy over time, the spectrum owner must utilize a spectrum broker. A **Spectrum Broker** is any system that can subdivide and assign a pool of spectrum rights to other users. For example, a broker could be everything from a human at a company selling allocations to third parties all the way down to an automated system assigning time domain slots in a multiple access system. This broker controls the dynamic policy of the spectrum owner to ensure that spectrum access rights are optimized according to the metrics set forth in that policy.

The rules set forth in the policy can use a number of different techniques to ensure that the spectrum is being used most efficiently at any given time. One use case described in IEEE 1900.4 [17] is **Dynamic Spectrum Assignment** which closely matches traditional assignment models, but allows the broker to change the assignment dynamically to meet policy goals in real time. Another use case is **Distributed Radio Resource Usage Optimization** which goes one-step further by having composite wireless networks exchange sensing and network information in real time to dynamically adjust the spectral resources between these networks in both time and frequency.

1.5.4.6 Environmental Knowledge

The ability to perform dynamic spectrum management could, in theory, rely entirely on negotiated agreements at different levels of abstraction such as between radios, between networks, and between spectrum owners. This would, however, leave inefficiencies in many locations. For example, noncollaborative incumbents that are sparsely distributed would still consume their entire allocation. In addition, changes in the interference environment or inconsistencies in the allocations could result in unusable or otherwise occupied channels being assigned to a new user. It is for this reason that the ability to sense the environment is critical for making decisions in a dynamic spectrum management scenario. For this reason, the 1900 standards treat network and environmental awareness as a critical enabling component to improving spectral efficiency [17].

In an IEEE 1900 system, the **Sensor** is any logical entity that provides information about the operating environment for a radio. This information is obtained through **Spectrum Sensing**, which is the act of measuring the RF environment for both detecting the presence of signals as well as characterizing those signals to determine how they behave. Information that can be provided by the sensors is broken up into two separate classes: *Sensing Information* acquired directly by the sensors about the environment and *Sensing Control Information* that provides metadata about the sensor itself. The first includes collected information from across the entire communications stack such as raw RF samples, a Clear Channel Assessment, data rates, the radio owner, and many more. The second category provides status and configuration about the sensor itself including any data archives or cognitive processes running on the sensor.

Data collected by the sensors can either be used locally on the radio node itself or, alternatively, provided to a central repository on the network. These repositories, called the **Data Archive**, are a place to store the distributed data from all sensors about the RF environment. This combined data allows Distributed Sensing, a method in which geographically separated nodes provide data to the archive to cooperate and collaborate on the total knowledge.

Similar to spectrum management, nodes that perform **Cooperative Sensing** are acting independently to achieve the common goal while **Collaborative Sensing** implies the nodes are actively communicating (perhaps through the data archive) to optimize the sensing. For example, a full picture of regional spectrum occupancy built through collaborative sensing would have the sensors communicating to fill holes in the sensed data while in the cooperative case sensors may simply do a random sweep the spectrum and report occupancy back to the data archive building a complete picture over time. This complete picture of spectrum occupancy and RF measurements over a regional area is referred to as an RF Environment Map that is combined with outside data such as geographic location, local regulations, and relevant policies to generate a full operating picture of the shared spectrum called the Radio Environment Map. This map, generated from sensor data and stored on the data archive, provides the complete picture necessary to make intelligent decisions about dynamic spectrum access in any given area.

1.6 Organization of this Work

Standards are written such that a device can be tested for conformity. Standards are not written to explain the theory behind the design decisions made in that standard.

The standards offer little to no justification for the limits imposed or choices made for that standard. One of the primary goals of this book is to elucidate wireless standards pertaining to wireless coexistence. To meet this goal, a theoretical background for the wireless coexistence must be provided. Those concepts can then be used to detail existing wireless coexistence standards including the motivation and tradeoffs surrounding different decisions made by the standards bodies.

The first portion of this book will largely follow the layers illustrated in Figure 1.6. Regulations will first be discussed, followed by a discussion on select concepts in communication theory relevant to physical signals. Those concepts will then be used to discuss initial contention mitigation among users equal in priority. This follows the regulation, to physical layer, to MAC layer paradigm in Figure 1.6. From there, more advanced concepts will be addressed detailing spectrum sensing and cognitive functions. These cognitive functions do not necessarily fit into a one dimensional protocol stack. As will be seen in Chapter 8, the IEEE 802.22 standard specifies a separate cognitive plane in its protocol stack to encapsulate these functions. Once the regulatory landscape, communication theory, and cognitive radio theory topics have been explored, the next chapters will detail the standards outlined in Section 1.4.

References

1 Newton, I. (1952). *Opticks*, fourthe, 1730. New York: Dover.

2 I. S. 1900.1-2019. IEEE Standard for Definitions and Concepts for Dynamic Spectrum Access: terminology relating to emerging wireless networks, system functionality, and spectrum management. *IEEE Std 1900.1-2019 (Revision of IEEE Std 1900.1-2008)*, vol. 23, 1–78 (23 April 2019). doi:10.1109/IEEESTD.2019.8694195. https://ieeexplore.ieee.org/document/8694195.

3 IEEE 802 (1999). *IEEE Std 802.11b*. IEEE.

4 Radio Frequency Devices Title 47 – Chapter I – Subchapter A – Part 15. [Online]. https://www.ecfr.gov/cgi-bin/text-idx? SID=e177016364e1b9e1b6c142ef6eb70fac&mc=true&tpl=/ecfrbrowse/Title47/ 47cfr15_main_02.tpl (accessed 9 October 2020).

5 Shannon, C.E. (1948). The mathematical theory of communication. *Bell System Technical Journal* 27 (4): 623–666. doi, https://doi.org/10.1002/j.1538-7305.1948. tb00917.x hdl:11858/00-001M-0000-002C-4314-2.

6 H. Nyquist, "Certain topics in telegraph transmission theory," *Transactions of the AIEE*, vol. 47, no. Apr, p. 617–644, 1928.

7 Zimmerman, H. (1980). OSI reference model--the ISO model of architecture for open systems interconnection. *IEEE Transactions on Communications* 28 (4): 425–432.

8 Voicu, A.M., Simić, L., and Petrova, M. (2018). Survey of spectrum sharing for inter-technology coexistence. *IEEE Communication Surveys and Tutorials* 21 (2): 1112–1144.

9 (2005). IEEE Standard for Information Technology – Local and Metropolitan Area Networks – Specific Requirements – Part 15.1a: Wireless Medium Access Control (MAC) and Physical Layer (PHY) Specifications for Wireless Personal Area Networks (WPAN). *IEEE Std 802.15.1-2005 (Revision of IEEE Std 802.15.1-2002)*, 1–700 (14 June 2005). doi:10.1109/IEEESTD.2005.96290. https://ieeexplore.ieee.org/document/1490827.

10 IEEE Standard for Information Technology – Local and Metropolitan Area Networks – Specific Requirements – Part 22: Cognitive Wireless RAN Medium Access Control (MAC) and Physical Layer (PHY) Specifications: Policies and Procedures for Operation in the TV Bands. *IEEE Std 802.22-2011*, 1–680 (1 July 2011). doi: 10.1109/IEEESTD.2011.5951707. https://ieeexplore.ieee.org/document/5951707.

11 (2013). IEEE Standard for Information Technology – Telecommunications and Information Exchange Between Systems. Local and Metropolitan Area Networks – Specific Requirements – Part 11: Wireless LAN Medium Access Control (MAC) and Physical Layer (PHY) Specifications – Amendment 4: Enhancements for Very High Throughput for Operation in Bands Below 6 GHz. *IEEE Std 802.11ac-2013 (Amendment to IEEE Std 802.11-2012, as amended by IEEE Std 802.11ae-2012, IEEE Std 802.11aa-2012, and IEEE Std 802.11ad-2012)*, 1–425 (18 December 2013). doi:10.1109/IEEESTD.2013.6687187. https://ieeexplore.ieee.org/document/6687187.

12 (2018). IEEE Standard for Information Technology – Telecommunications and Information Exchange Between Systems – Local and Metropolitan Area Networks – Specific Requirements – Part 19: Wireless Network Coexistence Methods. *IEEE Std 802.19.1-2018*, 1–458 (2 November 2018). doi:10.1109/IEEESTD.2018.8520953. https://ieeexplore.ieee.org/document/8520953.

13 (2003). IEEE Recommended Practice for Information Technology – Local and Metropolitan Area Networks – Specific Requirements – Part 15.2: Coexistence of Wireless Personal Area Networks with Other Wireless Devices Operating in Unlicensed Frequency Bands. *IEEE Std 802.15.2-2003*, 1–150 (28 August 2003). doi:10.1109/IEEESTD.2003.94386. https://ieeexplore.ieee.org/document/1237540.

14 (2019). IEEE Standard – Information Technology – Telecommunications and Information Exchange Between Systems – Wireless Regional Area Networks – Specific Requirements – Part 22: Cognitive Wireless RAN MAC and PHY Specifications: Policies and Procedures for Operation in the Bands that Allow Spectrum Sharing Where the Communications Devices May Opportunistically Operate in the Spectrum of Primary Service. *IEEE Std 802.22-*

2019 (Revision of IEEE Std 802.22-2011), 1–1465 (5 May 2020). doi:10.1109/IEEESTD.2020.9086951. https://ieeexplore.ieee.org/document/9086951.

15 Agreement Reached Regarding U.S. Position.(2003). https://www.ntia.doc.gov/legacy/ntiahome/press/2003/5gHzAgreement.htm (accessed 9 October 2020).

16 (2003). IEEE Standard for Information Technology – Local and Metropolitan Area Networks – Specific Requirements – Part 11: Wireless LAN Medium Access Control (MAC) and Physical Layer (PHY) Specifications – Spectrum ands Transmit Power Management Extensions in the 5 GHz Band in Europe. *IEEE Std 802.11h-2003 (Amendment to IEEE Std 802.11, 1999 Edn. (Reaff 2003))*, 1–75 (14 October 2003). doi:10.1109/IEEESTD.2003.94393. https://ieeexplore.ieee.org/document/1243739.

17 I. 1. S. Committee (2011). *IEEE 1900.4*. IEEE.

2

Regulation for Wireless Coexistence

Much in the same way that land, water, mineral, and other property rights are administered and regulated by a government, spectrum usage is similarly regulated throughout the majority of the world. Although one cannot perfectly apply the metaphor of physical property rights to spectrum occupancy, many similarities have arisen between the methods for regulating spectrum access the methods by which property rights are administered. In this chapter, we will discuss a brief history of how the shared resources of the spectrum have been managed, or not, by different governments as the use of wireless technology has grown and changed since its initial discovery. In particular, we will focus on the bands for unlicensed use as these provide an ideal testing ground for high occupancy spectrum sharing techniques.

2.1 Traditional Frequency Assignment

2.1.1 How Did It Work

Following the adoption of the more bandwidth efficient carrier wave transmission technologies, regulatory agencies in different countries were largely responsible for managing their own spectrum policies. Provided they operated within the loose framework adopted by the International Radiotelegraph Conferences and International Telecommunications Union, countries were free to determine how they wished to handle allocation of the *common use* spectrum resources in accordance with their respective political philosophies. For example, governments were free to sell *spectrum rights* to interested parties in the same way mineral and timber resources were managed. They might also allocate the spectrum resources to whomever was able to use that spectrum *best*, by whatever metric was used to

Wireless Coexistence: Standards, Challenges, and Intelligent Solutions, First Edition.
Daniel Chew, Andrew L. Adams, and Jason Uher.
© 2021 The Institute of Electrical and Electronics Engineers, Inc.
Published 2021 by John Wiley & Sons, Inc.

measure utility. If a country valued scientific research, then perhaps the spectrum most useful for sensing and radio astronomy would be assigned to public universities rather than broadcast radio operators. In reality, the vast majority of countries chose to adopt a hybrid plan, allocating spectrum to both commercial interests and public works through laws and regulations.

There are, in general, three paradigms for band assignments: exclusive band allocations, radio service allocations, and unlicensed allocations [1]. Exclusive band allocations grant a single entity (person, corporation, university, etc.) the right to exclusive access to that band for any use. Radio service allocations allow licensed users and equipment access to a specific band provided they follow a prescribed set of rules for that service. Finally, unlicensed band allocations typically require no license to transmit or operate equipment but are still regulated by a common set of rules.

Prior to the twenty-first century, commercial spectrum licenses around the world were almost always granted such that spectrum resources would be available for the sole and exclusive use of the licensee. This would mean that, provided owners operated within the band and power limits of their license, the owner could use that spectrum for any application at any time and expect to be protected from interference in that band. Interference, in this case, usually means man-made sources of noise and interference as opposed to natural phenomena. This type of interference can come from two potential sources, radiation generated as a side effect of operating machinery and intentional radiators that are incorrectly operating in a reserved band. Exclusive licenses are enforced through various consequences directed at people who, either intentionally or unintentionally, radiate in a reserved band, always requiring the offender to cease operation of the equipment that is interfering with the licensee and in the worst cases resulting in fines or imprisonment.

It is clear that allocating all spectrum resources using the exclusive access model is not acceptable. There are several use cases where granting access to the spectrum for arbitrary users is ideal and even sometimes required. For example, the very first cases of spectrum regulation were designed to ensure that maritime spectrum would be available for ships to communicate with one another for safety and navigation purposes. In this case it would not make sense to allocate the spectrum to a single shipping company, as the resource must be shared by everyone in the name of safety. A second use case is the *amateur radio operator*, hobbyists around the world who build radio systems designed to communicate across the globe with other radio enthusiasts for the sole purpose of entertainment. In this case, it would be wasteful to assign individual allocations to every single licensee for exclusive use. For these cases, entire bands can be allocated for a specific *use*. Often, this means that the *user* is licensed which grants them the legal authority to transmit in any of the bands set aside for this type of communications.

Early on, regulatory agencies recognized that certain types of industrial and consumer equipment would always generate large amounts of interference that could not be mitigated through other means. Regulators would set aside bands that would not be used for communication or sensing purposes and instead be relegated such that any device could generate interference in the band, but must also not be susceptible from interference generated by any other devices operating nearby. For example, one of the most commonly used unlicensed bands was first established at the 1947 ITU meeting in Atlantic City after concerns a newly developed technology that used microwave radiation to heat food would interfere with communications if they were allocated in the same band. This new allocation, the Industrial, Scientific, and Medical band, was dedicated such that "Radiocommunication services operating within those limits must accept any harmful interference that may be experienced from the operation of industrial, scientific and medical equipment" [2].

2.1.2 History of Allocations in the United States

Following the 1906 Berlin Convention, it quickly became clear that the federal government was going to have to a role in coordinating spectrum usage inside its own borders beyond simply complying with the Berlin Convention rules. The first official legislation from the US government relating to spectrum allocation came in the form of The Radio Act of 1912, officially called "An Act to Regulate Radio Communication" [3]. In this regulation, the three bands were allocated to the primary users of wireless technology: the US Navy, commercial shipping interests, and amateur enthusiasts experimenting with this nascent technology in their own homes. The US government was allocated the band from 187.5 to 500 kHz, commercial shipping interests were allowed to use the band between 500 kHz and 1 MHz, while the amateur radio operators were "restricted to a transmitting wave length not exceeding 200 meters and transformer input not exceeding 1 kilowatt" [3]. This effectively allocated the most useful bands at the time to the government and left the rest of the spectrum, everything above 1.5 MHz entirely open for experimentation. It should be noted, however, that this experimental band was not a free for all. Users were still required to apply for a license if they wanted to transmit in these bands.

Following the radio act of 1912 use of the spectrum for commercial purposes grew rapidly, especially the new field of "radio broadcasting." When the 1912 act was passed radio communications were almost entirely point to point voice conversations, a fact which is reflected clearly in the regulations used to allocate bands and license users set forth in the bill. As radio broadcasting became popular for distributing news and entertainment in a one-to-many format, the free form band assignments outlined in the 1912 act were shown to be ineffective.

Described as "chaos" [4], broadcast stations would move around within the band "at will" to get the best signal at a given time or location. In order to address the changing landscape of spectrum usage, the Radio Act of 1927 sought to increase the regulation of spectrum usage by creating the Federal Radio Commission, who would be charged "to regulate all forms of interstate and foreign radio transmissions and communications within the United States" [5]. This included the ability to perform spectrum allocation to different services, decide which license holders were allocated to which bands, and to regulate the types of content allowed to be broadcast. The primary work of the newly formed commission resulted in "General Order 40" [6], which created the AM Broadcast band for radio transmitters and allowed the commission to issue licenses to use one or more of the 96 *channels* between 550 kHz and 1.5 MHz. Channel licenses would be issued by the commission according to a complex set of rules that included considerations such as the geographical *zone*, the time of day, the allowed transmit power, the *equitable* sharing between states, and interference with neighboring assignments.

When it was first created, the intention of the FRC was that it would perform allocations and grant licenses and then be dissolved. The thinking at the time was that once licenses and band plans were in place, the Commerce Department could continue to manage spectrum as it had prior to the 1912 Act. However, turmoil surrounding the decisions made during the FRC's tenure and the number of legal challenges made following decisions it was clear that the Commerce Department could not adequately assign licenses full time [7]. For this reason, "The Communications Act of 1934" was adopted to provide a single governmental body for regulating the wireless spectrum, called the "Federal Communications Commission" (FCC) [8]. All of the radio regulation authorities that were previously distributed across the Commerce Department, the Federal Radio Commission, US Congress, and the Executive Branch would now be under control of the newly created FCC. These duties included allocation of the spectrum, granting of user licenses, device certification, setting penalties, and enforcing statutes all of which was now officially codified Title 47 of the United States Code [8]. It should be noted that Section 305 of the act does maintain the previous status quo of allowing the government control over their own allocations. This section maintains the idea that "the president," meaning the executive branch of the US government, holds the authority to regulate frequency assignments for federal usage [9]. This means that the newly created FCC would have authority only over the commercial and amateur use of the wireless spectrum, though it is clear that coordination between the two authorities would be necessary. Since 1978, the National Telecommunication and Information Administration (NTIA) has been the regulatory body charged with regulating federal use of the spectrum in conjunction with the FCC [10].

Prior to 1982, frequency allocations were assigned by the FCC through use of "comparative hearings." If more than one licensee was interested in a particular section of spectrum, a hearing was held to determine which applicant best fit the "public convenience, interest, and necessity" as mandated by the 1934 Radio Act [8]. In fact, this process was officially required following a supreme court decision in 1945 [11] that stated "Where the Federal Communications Commission has before it two applications for broadcasting permits which are mutually exclusive, it may not, in view of the provisions of the Act for a hearing where an application is not granted upon examination, exercise its statutory authority to grant an application upon examination without a hearing." Following this decision, each license granted was determined by the committee following a hearing that examined each applicant's case and determined which applicant would provide the best use of the band. These hearings were often expensive, both in terms of time and money, and often required three different rounds of hearings—one to determine if the applicants met requirements, one to review the applications themselves and determine how the licensees would use the band, and the final round in front of the commission itself [12].

Because the hearing process was onerous and time consuming, in 1984, the FCC began utilizing lotteries to assign the bands, starting with the first mobile telephony bands. Lotteries provided a benefit over the hearing model in the sense that the government no longer had to decide which applicant was *best*, a politically dangerous process, and that the assignments could happen at a very fast pace. While lotteries had the intended effect of speeding up allocations and removing potential bias from the allocations, there was an unintended side effect. Because anyone could apply for the license regardless of their ability to utilize it effectively, auctions turned into a speculative market in which middlemen would apply for licenses with the sole intention of selling their rights to the highest bidder on the open market [12]. This meant that the lottery system was, effectively, an auction system in which private parties reaped the benefits of a shared public resource. To combat this, congress passed the "Omnibus Budget Reconciliation Act" in 1993 that provided the FCC the authority to auction frequency allocations directly. This provides a hybrid approach to the previous methods in which licensing can be streamlined, but restrictions can be placed on the potential bidders to ensure the FCCs mission of spectrum utilization that benefits the public good. While these auctions suffer from some of the same issues that plagued the lottery system, the FCC is free to place obligations on the winners that allow license revocation if the bands are not used in the manner intended. For example, the auction that sold licenses in the 700 MHz for mobile telephony and data services required that winner provide coverage to "at least 35% of the geographic areas [...] within four years and [...] at least 70% of the geographic areas [...] at the end of their term" [13]. These types of restrictions would, theoretically, ensure that bands are used for their desired purpose and not purchased for speculative or anti-competitive reasons.

2.1.3 History of Spectrum Sharing

Even from the early days of wireless communications, it was clear that the concept of the spectrum as a shared natural resource was going to be important. Early spark gap transmitters consumed an enormous amount of bandwidth and the majority of early conferences on wireless telegraphy centered on how to ensure that critical ship to shore communications did not interfere with one another. The first International Radiotelegraph Conferences, held between 1903 and 1927, were focused primarily on ensuring radios could operate free from interference by establishing two sets of restrictions. First, the concept of frequency band allocations was introduced to ensure that transmissions which were very likely to interfere with one another were physically separated into different bands [14]. Second, the rules and procedures for transmitting on shared bands, referred to as protocol, were introduced [15]. These protocols established agreed upon ground rules for when different stations would have priority in shared bands and what types of communications those bands may carry. As technology progressed, these protocols became less and less significant. First, spark gap transmitters fell out of common use and were eventually banned in 1947 [2], freeing up a significant chunk of spectrum for the more efficient carrier wave systems. Second, the introduction of vacuum tube oscillators dramatically increased the achievable transmission frequency of carrier wave systems, thereby opening up a vast amount of additional spectrum for operators. Combined, these two events served to simplify the task of interference mitigation and *spectrum sharing* to simply allocating and assigning appropriate chunks of spectrum to users as described in Section 2.1.1. However, as the demand for mobile telephony and wireless networking technologies grew, this style of sharing became increasingly difficult. The next three sections will introduce how spectrum was shared following the initial Radiotelegraph conferences, how the demand for mobile phones spurred development in centrally managed spectrum sharing, and how wireless networking technologies changed the way unlicensed spectrum is managed.

2.1.3.1 Pre "Mobile Phone" Era (Easy Mode!)

During the earliest days of radio, referred to "wireless telegraphy" at the time [16], the common technology for broadcasting utilized Franco Marconi's *spark gap* transmitters. Because this style of transmitter generated what are called *damped waves*, transmissions consumed a very large amount of bandwidth and therefore interference between different transmitters was very common. For these reasons, there was significant concern among navies and shipping companies alike that the wireless telegraph would be rendered useless through a wild-west style **tragedy of the commons** where their own transmissions would be constantly interfered with by other transmitting stations. To address these concerns, the first intercontinental discussion on radio regulations was held in Berlin in 1903. The meeting, dubbed the "The First International Radio Telegraphic Conference," was a discussion between eight countries with the goal of ensuring that the navies and shipping interests of these nations could communicate at sea with wireless telegraph equipment [16]. Following the conclusion of this conference, the representatives from each attending country agreed to present an agreed upon protocol for radio etiquette that ensured fair use of the airwaves. In the following years, the spark gap transmitter technology was largely abandoned in favor of the newer, and much more bandwidth efficient, *carrier wave* (CW) technology. This new technology allowed transmitters to use significantly less bandwidth and largely eliminated the interference concerns associated with the spark gap transmitters. However, the precedent for international coordination of the shared spectrum resources was set. Countries around the world began establishing their own organizations to manage spectrum within their borders in compliance with the various international agreements in place. At the same time, organizations such as the International Telecommunications Union and International Radiotelegraph Conference continued to foster international cooperation and ensure a certain degree of compatibility between radio licensing systems of different nations.

Following the initial flurry of international coordination of spectrum there was a long period of relatively insular growth in wireless broadcast technologies. With global adoption of the more efficient carrier-wave broadcast technologies and international bans on high power spark gap transmitters the wireless spectrum was effectively a wide-open space that required only loose coordination between geographically disparate countries. While bands were carved out for different uses and codified through the various ITU and IRC agreements [17], there was still significant leeway granted to individual countries regarding how those bands would be licensed and allocated within their own borders.

2.1.4 Mobile Phone Explosion

Prior to the 1980's, wireless communications operated on either a one-to-one medium such as telegraph and telephone conversations or a one-to-many medium such as television and radio. In both cases, the connection between the communicating parties was limited by the range at which the users were able to transmit and receive effectively with a single radio on a single frequency. With the rise of digital computers and digital networks connecting them, machines could now be used to control access to the wireless spectrum dynamically. Spectrum allocations no longer had to be static and time invariant. Machines could be programmed to only use the wireless spectrum only at certain and predictable times coordinated through prearranged protocols. While this type of automatic control made first wireless computer networks such as ALOHAnet [18] possible, it first found use in the rapidly developing field of mobile telephony. Although commercially available "mobile telephones" had existed since as early as 1949, services were essentially operator coordinated point-to-point radio links. Radios were assigned a fixed frequency and their receivers were patched into the local telephone network manually by an operator at first and later with automated dialing similar to that used with landline telephones. Even the first generation cellular mobile telephony systems, such as AT&T's Advanced Mobile Phone System (AMPS), required a single frequency assigned to each user. The breakthrough technology in these "first generation" systems was the automatic process by which frequencies were dynamically assigned to users when they wanted to make a call. Starting with the "2nd generation" (2G) mobile telephony services standards such as the Global System for Mobile Communications (GSM) and Interim Standard 95 (IS-95) [19] were able to use computer controlled networks to share the same frequencies in the spectrum among a large number of users with various multiple access technologies. These technologies and standards have proved to be the initial model for what we today call **Spectrum Sharing**. Rather than a dedicated frequency assignment model such as those used in 1G cellular and "0G" point-to-point systems, these new standards sought to facilitate real-time, automated sharing of the spectrum between multiple users. These 2G standards, and the later generation of mobile data and telephony services, generally facilitate sharing through a model of centralized control. The individual regional areas, called cells, are controlled by a single base station that tells all the other radios in the area how and when to share the spectrum. This centralized model of control provides a very efficient use of spectrum at the cost of compatibility. Anyone who does not speak the same protocols or conform to the central controller is unable to communicate.

2.1.5 Wireless Networking

Alongside mobile telephony, a rapidly growing application for spectrum resources is private wireless networking. Heavily utilizing unlicensed spectrum, technologies such as WiFi, Bluetooth, and other low power networking protocols have exploded in use since the late 1990's [20]. Initial wireless networking standards were designed to replace relatively short-range cables, such as LAN wiring in the home, RS232 serial cables for computer peripherals, and audio cables for headphones. Unlike mobile telephony, where deployments are managed by large corporations, these particular technologies were mostly operated in homes and small businesses necessitating a very different model for spectrum assignment. Because it would not be feasible for consumers to acquire their own spectrum licenses from the government in order to operate a wireless LAN, the initial protocols relied heavily on use of unlicensed spectrum in the 2.4 GHz **Industrial, Scientific, and Medical** (ISM) band. At the time, this band was attractive for two primary reasons. First, this section of the ISM band was harmonized, meaning that it overlapped all three ITU regions and ensured that devices could be used worldwide. Second, this band provided a good trade-off between range and equipment costs. While using the higher frequency harmonized allocations would reduce the probability of interference between different deployments, it also meant more costly precision RF components. Therefore, in the early days of wireless networking, the 2.4 GHz allocation provided a good balance between propagation range and component costs that made it attractive for use in consumer devices [21].

Since the introduction of these original *cable replacement* technologies, the scope of wireless networking technologies has expanded dramatically. Growing beyond simple cable replacement mechanisms, standard implementations now cover a wide variety of geographic coverage ranges, network interconnection models, licensed and unlicensed frequency ranges, and user allocation paradigms. For example, geographic coverage of networks can range from citywide *last mile* networks that provide internet access to homes down to *body area networks* that connect sensors and devices worn by a single person. In addition, these technologies are finding use outside the traditional network access roles in areas such as security monitoring equipment, sensor data collection, home automation, smart city management, and many more.

2.1.6 Future Allocations for Coexistence

Until this point, frequency assignments have been treated as an *all or nothing* proposition. Either license holders can be allocated a band to do with as they please, or the band is open to everyone for anyone to use with no guarantees on availability. There is another model, however, that can be applied and provide the best of both worlds: **Intelligent Spectrum Sharing**. While static frequency allocation may technically meet the definition of *wireless coexistence*, it is naive at best and results in a significant amount of unused or underutilized spectrum [22].

It is clear that, moving forward, employing intelligent coexistence strategies will be required to meet the growing spectrum demands of mobile data services, wireless networking technologies, and smart connected cities. What is less clear, however, is a path that moves the world from the existing monolithic frequency assignment model to one of intelligent coexistence across the entire spectrum. In general, there are three main considerations when it comes to gradually adopting smart coexistence strategies: preventing primary user interference, guaranteeing fairness for the public good, and ensuring commercial access to bands.

The first consideration, preventing **Primary User** (PU) interference, is critical to ensuring that legacy users of spectrum are not negatively affected when they do not participate in smart coexistence strategies. Because these legacy users potentially have a significant capital outlay in dumb equipment that assumes access to the spectrum at all times, it may not be feasible to simply implement a smart sharing scheme. The legacy devices do not speak this sharing new scheme and will therefore be unable to participate in a meaningful way. This means that the **Secondary Users** (SU) seeking to occupy underutilized spectrum will be forced to simply move to avoid interfering with the PU. While this ensures that the PU of the channel is protected, it also means that large swaths of spectrum could remain underutilized in an effort to ensure legacy devices are protected. For example, as the lower frequency bands fall out of favor for satellite communications, ground based terminals could begin using that spectrum. However, there it would be very difficult for the mobile SUs to determine if there were any receiving satellite stations nearby. Solutions must be researched that allow effective sharing of the band by giving SUs the information they need to know when and where they are allowed to operate without interfering. This may include new sensing technologies for detecting active PUs, shared databases that provide channel information to users, or a completely new alternative that ensures the PU is protected while allowing the SU to access the spectrum.

Second, it will become clear in the next section that spectrum resources are a public resource and, as such, must be shared with everyone to maximize the greater good. While it is clearly difficult to determine metrics by which to measure *the greater good*, it is the responsibility of the government to ensure that spectrum is used both fairly and economically such that all citizens receive the benefit of the resource. While this could certainly mean only providing dedicated licenses to commercial entities, it is far more likely that a hybrid approach to allocation that balances private and public use would result in the best benefit to the citizens.

Finally, it is clear that if spectrum access is to be regulated differently to ensure access for secondary users then there must still be some spectrum set aside for complete commercial control. There are certainly some cases where it does not make sense to allow secondary users access to a band at all. Services that require low channel occupancy but a high reliability would be dramatically affected by SUs employing a spectrum sensing strategy. For example, a security company that employs wireless detection sensors might only use a small portion of the allocated bandwidth during normal operations. If a SU were to occupy the channel believing it to be vacant, that SU may interfere with a critical alarm update if the legacy system does not perform proper channel assessment because it believes it has sole access to the system.

It will be up to regulators to allocate spectrum between the traditional monolithic assignments, the fully unlicensed bands, and the intelligent sharing schemes in a way that ensures the interests of the public are fully protected. This could mean selling off large chunks of the spectrum to generate funds for public programs or it could mean allocating large chunks solely to common-use protocols that are spectrally efficient. The more likely answer is that a thoughtful approach to resource management that balances the potential revenue from auctions with the intrinsic value to the people of freely available spectrum.

2.2 Policies and Regulations

2.2.1 Spectrum Rights and Digital Commons

One of the fundamental questions that must be answered when discussing policies and regulations regarding spectrum usage is a very simple question—"Is spectrum a natural resource?" Whether the spectrum is considered a natural resource drives an entire line of questioning about the role of governments in its management. For example, should spectrum be treated in the same way that forests and minerals are treated? Should it be subdivided and owned in the same way land is granted? Is it something that must be protected and cared for deliberately in the same fashion as critical waterways? These are questions that economists and legal experts have been asking themselves since the 1950s [23]. The early days of spectrum regulation were not concerned with the idea of property rights, the public good, legal questions about who *owned* the spectrum, or even questions about if it could be owned at all. Initial spectrum regulation treaties were focused on solving an engineering problem, not a legal one. Through international agreement, everyone could share the resource and ensure that everyone had access when they needed it. As the adoption of additional wireless technologies picked up steam and regulators had to begin deciding how spectrum should be allocated between different applicants, the questions of how and if the government should be assigning these "property rights" started come in to play [23].

The idea of property rights, and the answer to whether or not something can be *owned*, is controlled by a number of fundamental legal factors [24]. Speaking in the abstract, these different attributes can help people frame the problem in a way that can help determine if something can be owned and, if it can, if it should be considered a *public resource* to be managed by governments in the interest of the greater good. While the answer to the latter question will depend heavily on political ideologies, the former can be used to help determine the important factors at play. The first question, "what makes something property?" is answered by examining the *bundle of rights* that a user has. The different properties in the bundle include the concepts of free use, exclusivity, and transferability [25]. The first right, **Free Use**, dictates whether the user is able to use the resources in any way they see fit.

Second, is whether the user or owner is granted **Exclusive** rights to access (or choose who can access) the resource in question. If the owner is able to control who is allowed to use the resource and when they are allowed to do it, that user is said to have the right of exclusivity. Finally, the idea of **Transferability** dictates that to be property, it must be able to be transferred between private parties with minimal involvement from the granting government. When viewed through this lens, the question of whether or not spectrum should be considered property is clear: spectrum is property [26].

Given that spectrum is considered property, the next question that must be asked is whether it is a common property that must be shared among the people or whether permanent licenses may be granted. In the United States, for example, the 1934 Communications Act effectively banned private ownership of spectrum, citing its status as a public resource that must be managed by the government for the public good. This notion, however, is not a simple on its face as "spectrum is reserved for the people." Like other resources reserved for the common good, the government is free to grant licenses as it sees fit to maximize the public good. This could mean granting exclusive rights to bands through auctions as a way of raising capital for domestic social programs. In this case, a "license to operate" is nearly indistinguishable from *owning the spectrum*[27]. A license holder is able to exercise the same *bundle of rights* that a property owner could after exchanging something of value (i.e., money) for those rights. Provided that license terms are not in direct opposition to the greater good, this model of auctioning spectrum rights to the highest bidder should result in the most equitable benefit to the people [26].

While a full discussion on the concepts of common property rights is outside the scope of this book, this section should have provided a sufficient understanding of the primary arguments for and against centralized allocation of shared resources such as spectrum. These issues provide a fundamental basis for answering questions about if, how, and why spectrum use should be shared as a common resource for all people. As regulatory agencies move away from the traditional property rights model these questions of whether spectrum should be designated *for the greater good* drive the technical discussions about how and when spectrum can be shared among different users. For example, it is clear that for legacy license holders the amount of interference from spectrum sharing technologies they would prefer is *none*. However, it has been deemed that maximizing the use of these shared resources is in the best interest of the public. Therefore, the regulatory agencies must demand a compromise that allows for some small probability of interference in exchange for the ability to maximally use spectrum that license holders leave fallow. Weighing the impact of spectrum sharing to primary license holders against the net gain in *greater good* is not an easy question to answer and the legal theory regarding rights to natural resources and other types of common property will ultimately provide input to the question of this balance.

2.2.1.1 The Case Against Application of "Common Pool Resources" Economics

Resource allocation in the electromagnetic spectrum is often compared to allocation of land and, for this reason, laws and ideas surrounding land usage and ownership are regularly applied to the spectrum [28]. While there are several properties in common between spectrum use and land use, there are also some key differences that make it difficult to apply the logic of the law directly. One of the biggest questions is to first determine if the EM spectrum is, in fact, a resource that should be shared and managed as opposed to owned outright.

One of the primary points that makes electromagnetic spectrum different from other natural resource arguments is that it is *reusable*. While using the spectrum at any given time or place "consumes" that chunk of spectrum, it is available for someone else as soon as it is done being used. This is in direct opposition to traditional natural resources such as forest providing lumber, a field providing nutrients for crops, or a mine providing ore. In the traditional cases, once something is allocated and consumed it is no longer available and, therefore, no longer valuable. This key difference makes it very difficult to apply traditional economic modeling to spectrum use.

Given the differences between electromagnetic spectrum and the traditional concept of natural resources, it is tempting to apply the concept of land and property ownership to spectrum instead. While this model is closer in the sense that the resource can be reused over and over this remains an imperfect analogy. Unlike land, spectrum cannot be "polluted" in the sense that can be ruined forever. The primary cause of spectral pollution is other man made devices transmitting. Even if the task is daunting, it will always be possible to clean up any section of the spectrum by decommissioning radio services causing interference. This prevents applying *tragedy of the commons*[29] thinking. Research centering on the commons relies on the fact that a shared resource can be reasonably used by individuals and yet overused in aggregate which, in turn, makes it unusable for everyone.

The fact that it is nearly impossible to permanently consume or pollute the wireless spectrum makes it difficult to apply traditional economic thinking on how this shared resource should be used for the greatest benefit.

Occupying spectrum is analogous to occupying a particular space on earth in the sense that once you have vacated that space, it is now free for someone else to use.

2.2.1.2 What to Do Then?

There are several common proposals for consolidating underutilized spectrum. These include decommissioning the old services if they are of limited utility, moving the existing incumbents into newly organized bands that optimally fill the frequency resource map, and allowing for secondary and unlicensed users to operate in unoccupied bands provided they do not interfere with the allocated Primary User.

Given that spectrum is a common resource, it is clear that if someone is not using their spectrum allocation to its full potential a case could be made that it is reasonable to alter the terms of the license for the greater good under the principle of eminent domain.

2.2.2 Spectrum Coordination (Both Licensed and Unlicensed)

2.2.2.1 National and International Regulatory Bodies

Even from the early days of radio, it was clear that international cooperation was going to be required to prevent interference between operations in neighboring countries. At the first radio conference in 1903 the major powers experimenting with radio at the time came to the conclusion that because it was impossible to stop a physical wireless transmission at a political border, coordination would have to occur in order to ensure that radios could operate seamlessly near the borders of countries with disparate regulations. The documents from the early meetings [14, 30] are all very clear about the fact that the sovereignty of the individual countries is first, and foremost, the driving factor in spectrum regulations but recognizes that through harmonization of bands everyone will be able to use the bands efficiently. In fact, the driving idea behind the radio conferences was to generate a minimal set of rules that would ensure interference free operation while allowing the individual countries to control the majority of spectrum allocations within their own borders. From these original meetings, several international regulatory bodies grew up in the disparate geographic areas of the world. In these early days of international cooperation, the primary focus was on preventing interference between when radios built to a specification in one country were brought near the borders of another.

Global regulations and agreements, such as those defined by the ITU and at the International Radiotelegraph Conferences were, in large part, focused on maritime and aviation related issues. This led to a particular set of restrictions on the member countries that left significant band areas open for general use. Neighboring countries, however, have the same problem of cross-border interference. Because this interference can be from permanent, fixed transmitters it has the potential to be even more harmful than interference from ships and planes. For this reason, a number of regional regulatory agencies were created through treaties and agreements to set further restrictions and harmonize band allocations for fixed transmitters near political boundaries.

The ITU-R is the largest and most comprehensive spectrum regulatory body in the world. At the time of writing, there are 193 member countries of the ITU of the 195 countries in the world [31]. With this global reach, the ITU-R is able to set

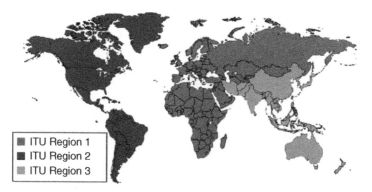

Figure 2.1 ITU regions.

basic regulations to ensure that all member countries can freely operate wireless devices on their sovereign soil without interfering with other countries and without fear that other countries will interfere with them. The primary duties of the ITU-R are to ensure equitable access to spectrum, protect frequencies designated for international distress calls and safety equipment from interference, and mediate cases of harmful interference between member states [32]. In order to simplify the application of regulations, the ITU-R recognizes three primary geographic regions grouped by continent. Shown in Figure 2.1, Region 1 covers Europe, Africa, and the former USSR portions of Asia, Region 2 covers the Americas, and Region 3 comprises the majority of Asia and Australia.

Within the Radio Regulations, the ITU harmonizes frequency allocations according to these regions. When possible, best efforts are made to harmonize across regions as well, providing bands that ensure interference free operation across the globe. This harmonization, both regional and global, provides two main benefits. First, it achieves the primary goals of the ITU in eliminating harmful interference in the spectrum across borders. Second, it allows for equipment manufacturers to create products that can be used and sold globally without requiring changes to the hardware. By harmonizing the radio bands, adoption of new radio technologies can be sped up dramatically when only one radio must be designed, tested, and certified.

Today, the primary avenue for achieving the goals of the ITU-R is international coordination that occurs at a series of meetings called the World Radiocommunication Conferences (WRCs). Held every three years, the attendees of the WRC meet to discuss "any radiocommunication matter of worldwide character" [33]. Activities at these conferences include updates to the radio regulations based on a combination of political changes and scientific research completed since the last conference. The result of these meetings is an update to the "Radio

Regulations" (ITU RR) [32], which is a list of legal articles that describe how and when different countries may use different frequency bands. Originally, these regulations were generally very broad leaving the member states to divide up the bands within their own borders as they saw fit. However, in an increasingly globalized world economy, the desire for equipment interoperability around the world has led to a number of harmonization efforts that add increasingly stringent requirements on certain bands.

The shortest path to the adoption of wireless coexistence technologies lies with the ITU Radio Regulations. If the ITU members can agree on a harmonized band that provides the regulatory framework needed to allow smart spectrum sharing, then global adoption of these technologies will certainly occur. With a set frequency band and clear rules on use, research scientists and equipment manufacturers can quickly iterate toward solutions that make optimal use of the spectrum for the greater good. As a first step, dedicating a small section of spectrum globally to the effort would certainly fall into the ITU's mission "to facilitate equitable access to and rational use of the natural resources of the radio-frequency spectrum"[32].

Outside of the ITU, there are a number of regional spectrum regulatory agencies that seek to fulfill the same mission of harmonization, but at a smaller scale. These bodies are typically more focused on local issues in a region such cross-border interference and ensuring equitable access to the spectrum for everyone. These include the Inter-American Telecommunication Commission (CITEL), the European Conference of Postal and Telecommunications Administrations (CEPT), the Asia-Pacific Telecommunity, and the African Telecommunications Union. Figure 2.2 shows the international coverage of these bodies respectively. Through

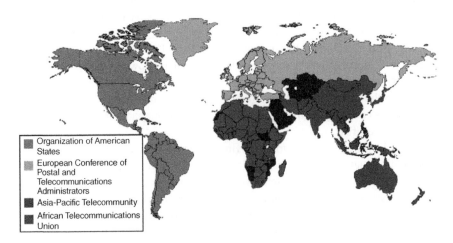

Figure 2.2 Coverage of regional spectrum authorities.

a combination of the ITU and these regional authorities, it can be ensured that the shared spectrum resource is available to all.

The Inter-American Telecommunication Commission (CITEL) is part of the Organization of American States (OAS), founded in 1948, which sought to establish "an order of peace and justice, to promote their solidarity, to strengthen their collaboration, and to defend their sovereignty, their territorial integrity, and their independence" [34]. Given the objectives of the OAL, CITEL is tasked with two primary functions. First, to promote harmonization of technical standards across the member countries ensuring interoperability and eliminating harmful interference. The second goal is to promote the development and use of technologies that improve the communications ability in "underserved areas."

The Electronic Communications Committee (ECC) of the European Conference of Postal and Telecommunications Administrators (CEPT) "considers and develops policies on electronic communication activities in European context" [35]. Member countries are focused on improving spectrum sharing and ensuring that local regulations allow for the adoption of new technologies without interfering with legacy users.

The Asia-Pacific Telecommunity (APT) was founded in 1979 is focused on governmental cooperation with equipment manufacturers and research institutions to ensure "qualitative and sustainable growth" of communications infrastructure in the Asia-Pacific Region [36]. The objectives of the APT are to expand telecommunications services and information infrastructure to the maximum benefit of the people in the region.

The African Telecommunications Union (ATU), founded in 1977 as part of the Organization of African Unity, seeks to improve access to information infrastructure and services throughout the continent. The end goal is to provide universal network access across the continent as a means of improving the standard of living for its citizens [37]. Its member states participate in international regulatory activities with goal of ensuring that Africa's interests are promoted. They believe that Africa provides a unique environment from the rest of the world and seek to ensure that their ability to quickly deploy new infrastructure across a wide, sparsely populated geographical area is not hindered by regulatory decisions focused on the dense urban deployments of Europe, Asia, or the Americas.

2.2.2.2 Ensuring Compliance Through Certification and Licensing

When considering how unlicensed bands would be regulated, there are a number of considerations that regulatory bodies must take into account. First and foremost, it is clear that just because a band is "unlicensed" does not mean that it is completely without regulation. First and foremost, the regulatory agencies are tasked with ensuring that there is no interference to the authorized users of allocated bands. This means that they must, by definition, place restrictions on how adjacent unlicensed bands may be used if only to ensure that nearby dedicated allocations are not affected. When considering the use of telecommunications equipment in any band, national regulators generally have the option of certifying either the operator, the equipment, or both.

In the case of equipment regulations, local laws could prevent the sale or importation of equipment that has not passed testing requirements that verify compliance with a band's restrictions. For example, certifying a maritime radar for sale might involve the regulator verifying that the user is restricted to only transmitting on the allowed frequency and that the unit does not generate harmful interference in other bands. This provides assurance to the regulator that even in the hands of an untrained or malicious user, the device is physically incapable of generating interference to other users of the band. This is an especially useful tool for bands where it is expected that the users will have little or no knowledge of the laws and regulations regarding transmitting such as in the mobile telephony or personal radio service bands.

In some cases, it may not make sense to restrict the operation of equipment provided that there is liability that can be assigned to the operator if interference is generated. In this case, the onus for ensuring that radio transmissions comply with local regulation is on the user of the equipment rather than on the manufacturer. This type of model makes sense in scenarios where equipment needs to be quickly easily reconfigurable but also has a very wide range of configuration options. For example, amateur radio services have a wide variety of modes and modulations that are available to the user which are most efficient at different times and places depending on the atmospheric conditions at any given point in time. It would not make sense in this case to restrict the use of different bands to specific configurations as it would make communications difficult as band conditions change. In this case, by licensing the operator and not the equipment, the user is able to effectively use the spectrum in the most efficient way possible even under rapidly changing spectrum conditions. This flexibility comes with the understanding that while a user could generate interference to another allocation, either intentionally or unintentionally, it would not be in their best interest to do so because they could potentially face fines, the loss of their license, or in some cases jail time.

Realistically, regulatory agencies do not have to choose one model or the other. Some combination of both user and device certification is required for almost any radio sold in the world today. For example, devices sold in the US must comply with the FCC's part 15 regulations that limit the amount of energy the device can emit both intentionally, for communication purposes, and unintentionally as a byproduct of its operation. Figure 2.3 provides examples of different types of radio services that have differing levels of restriction on the user versus the equipment.

	User heavy restricted	User lightly restricted
Equipment heavily restricted	• Cellular Telephone base stations • Fixed Wireless • Internet Services • Provides • Satellite • Communications • Equipment	• 802.11 • Networking • PRS "walkie-talkies" • CB Radio base stations
Equipment lightly restricted	• Amateur Radio base stations • Mobile Maritime • Radios	• Radio control cars • Personal FM Radio transmitters

Figure 2.3 Radio service restriction matrix.

2.2.3 Case Study in Spectrum Reallocation

Consider the case of an allocation around 2.2 GHz in use by several federal agencies following the passage of the National Broadband Initiative which was, at the time, allocated globally toward for use in Space Research (SR), Space Operations (SO), and Earth Exploration Satellite Services (EESS) [38].

This band in particular has a number of properties that make it an excellent case study for exploring the concepts that affect the decision to re-allocate spectrum assignments. The physical properties of the channel, the incumbent users and their applications, and the difficulty of deploying new radios all have an impact when evaluating spectrum for re-allocation.

First, this particular frequency band was chosen because it provides a good tradeoff for the communication subsystems in space vehicles. The wavelengths in use are short enough to allow small form factor antennas while simultaneously providing good atmospheric propagation characteristics in all weather conditions. This allows spacecraft designers to utilize a single antenna for both regular and emergency operating conditions as opposed to a more traditional model requiring a high gain horn for day to day operations with an omni-directional backup for emergency situations in which the satellite is unresponsive. Ultimately, because these properties are a function of the frequency used for communications, they are not something that can be easily moved or reallocated. If the allocation were moved higher, atmospheric conditions would require the transmitter to use more power for completing the link in bad weather and potentially require directional antennas that would need to track the earth's surface. Conversely, if it were moved lower, the size and weight of the spacecraft antenna would be increased, affecting the overall cost to deploy and run the satellite.

Second, at the time of the NTIA study this band was in use by a number of government entities for both command and control (C2) and wideband links for transferring research data. The users included the Hubble Telescope, NASA's Tracking and Data Relay Satellite System (TDRS), the Air Force's Space Ground Link Subsystem (SGLS), and a host of other short term operations such as terrestrial monitoring of aircraft, flight testing of missiles, and point to point microwave relays [39]. This particular intersection of users makes reallocation difficult due to the fact that they are all providing mission critical support in the protection of the United States. Extensive research and development has gone into the equipment in order to ensure that they meet the exacting standards required by scientific and military users. If these users were moved, the compliance and performance testing for these mission critical components would have to be repeated. Consider the use cases of this band in comparison to something like the push-to-talk voice networks utilized by the National Park Service or wireless sensor networks used to monitor crop health by the US Department of Agriculture. While the latter use cases are certainly important, they would be significantly less impacted by an unforeseen complication in new radios than the US Air Force or NASA Mission Control.

Finally, this band represents the "worst case" when it comes to considering the replacement cost of potential radios. Spacecraft currently using this band cannot feasibly be replaced or upgraded. This implies that any increase in the amount of interference or a reallocation of the band would result in decommissioning every spacecraft designed to use this band. Compare this to the Digital Broadcast Television transition in the US [40]. That endeavor meant replacing or retrofitting every TV in America. While that is a daunting task to be sure, it is nothing compared to sending technicians to the outer reaches of the solar system to replace a radio on the Voyager [41] spacecraft!

As described in Section 2.1, the common methods for freeing up spectrum would include decommissioning the incumbents, moving them to a new allocation, or allowing secondary users on a noninterfering basis.

The first method to support freeing up the band, decommissioning the incumbent services is extremely unlikely. With several national governments utilizing this spectrum for space operations, it is unlikely the US would be able to convince their international partners that this move would be worth the cost. In addition, the US itself relies heavily on this band for protecting its own interests.

Examining the second method, re-allocation, it is clear from the start that this particular case would make that very difficult. First, the nature of the currently deployed networks ensures that the reallocation would have to happen globally due to the fact that spacecraft often orbit the earth and would interfere with any geographic region not participating in the re-allocation. Second, because the deployed equipment is so difficult to reach, it would require replacing every spacecraft currently using this band at monumental cost to the governments around the world. The one realistic option for reallocating the band would be to, in one sense, place a *geographic restriction* on the use of the band. If the incumbent users of the band were able to restrict their use exclusively to space, ground users would be free to transmit and receive without interfering with the operation of the satellites. In this scenario, the command and control radios of the satellites would be offloaded to relay satellites located in geosynchronous orbit. These relay spacecraft would translate transmissions from ground stations at a different frequency allocation to the old 2.2 GHz band and relay them to the legacy satellites as shown in Figure 2.4. However, this solution only alleviates the problem of interference with the ground stations. If the new users of the band were allowed to transmit with sufficient power, they may still interfere with the spacecraft itself. For this particular case, reallocation of the band does not make sense due to the enormous cost of either radio replacement or mitigation.

The final potential method for freeing up this band, allowing secondary users to occupy the underutilized spectrum resources, is the most likely candidate for this use case. In this case, there are a number of considerations that need to be taken

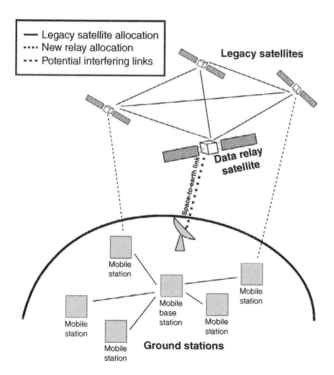

Figure 2.4   Protected space relay reallocation concept.

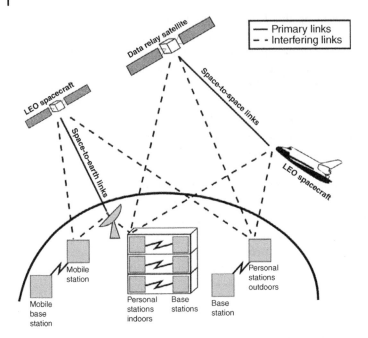

Figure 2.5 Potential interference sources for vulnerable space services. *Source*: Adapted from Rec. ITU-R SA.1154 [38].

into account when analyzing the impact that new terrestrial users would have on this band. Significant research was previously undertaken by NASA and the ITU to determine both the likelihood and severity of interference from ground users operating in this band [42]. Due to the relatively low power levels involved, the radios in this band are highly susceptible to interference. Therefore, each phase of the relayed, bi-directional, radio link must be considered separately to determine the potential impact. Figure 2.5 shows an overview of the operations that may happen in this band. Interference from ground users will manifest itself separately depending on which path the communication is using. Utilizing the system parameters for the proposed IMT-2000 standard, the ITU report demonstrates the potential impact that sharing the band would have on the spacecraft communication links. Analysis was performed separately for the potential uplink (earth-to-space), downlink (space-to-earth), and cross link (space-to-space) scenarios and found that the overall impact to spacecraft would be too great to make the additional terrestrial bandwidth worthwhile.

With the difficulties presented by this band, it is not a surprise that the NTIA ultimately removed this band from consideration. Utilizing the existing research performed by the ITU, which indicated that reusing this band for ground based networks would cause significant interference to the space vehicles, the NTIA determined that the operations could be neither discontinued, moved to other regions of the spectrum, nor accessed by Secondary Users in a noninterfering fashion [43].

2.3 Bands for Unlicensed Use

2.3.1 Overview of Unlicensed Use

One common misconception when it comes to unlicensed band usage around the world is that operating in allocated bands is a completely unregulated affair that allows radio operators to transmit at will with no regard for the consequences. This misconception stems from the idea that because unintentional radiators were the primary motivation for allocating these bands, and they are by definition unintentional and difficult to control, and that the same rules apply to intentional radiators operating in this band. In fact, the exact opposite is true: Article 1.15 of the ITU Radio Regulations reads:

> 1.15 industrial, scientific and medical (ISM) applications (of radio frequency energy): Operation of equipment or appliances designed to generate and use locally radio frequency energy for industrial, scientific, medical, domestic or similar purposes, excluding applications in the field of telecommunications.

The key point of this language is the final words, "...excluding applications in the field of telecommunications." This means that, in fact, the operation of unlicensed communication devices in the allocated bands are controlled via the ITU and its member countries are regulated in a variety of ways. When ITU member countries generate their National Table of Frequency Allocations (NTFA), they must provide a Generic Use Authorization (GUA) for bands that allow unlicensed communication devices [44]. These generic authorizations provide a number of restrictions on the types of devices that can be used for communication in the unlicensed bands including both technical and operational requirements established to ensure that these devices will not interfere with licensed users [45]. Today, the more specific term "license exempt" is often used by international regulators as it provides a clearer distinction on the regulated status of the devices. License exempt devices are those that operate with a non-interference basis according to the local GUA and are subject to interference from other devices operating in the same band.

2.3.2 Voice and Other Restricted but Unlicensed Bands

While not strictly "unlicensed," there are a number of bands designated by the ITU and national regulators as either "license exempt" or "licensed by rule" [46]. These particular styles of spectrum management can be thought of as a middle ground between single user licensing and unlicensed spectrum. In these styles, individual users are not required to obtain a license to transmit or otherwise use the band, but they often must comply with strict regulatory requirements on their use of the band. Additionally, for a "licensed by rule" band, the user is required to obtain a blanket license to operate. The justification for this style of licensing comes from a desire to reduce the burden on regulatory agencies while introducing a minimal amount of risk that the users of these bands could introduce harmful interference to other licensed services [47]. The general idea of the blanket license is that user equipment would be sufficiently low power or otherwise restricted that even in the hands of an unskilled operator, the equipment would be unlikely to cause interference. This alleviates the need for regulatory agencies to process the very large number of license applications that would come from users in this band and frees up resources for other needs.

A prime example of a "License Exempt" scheme are the various "Personal Radio Service" allocations around the world. These allocations are designed to allow citizens to communicate with minimal licensing requirements for light duty personal use. Typically, PRS radios are either hand held or vehicle mounted mobile radios that operate on a fixed set of frequency channels and are limited by design to a set maximum transmit power. Examples of a personal radio service include the "Citizens Band" (CB) and "Family Radio Service" (FRS) radios in the United States or the 446 MHz Private Mobile Radio (PMR446) service in the European Union. All of these allocations allow users to buy Commercial-Off-the-Shelf (COTS) radios and begin transmitting immediately without any additional licensing. However, while the user does not require a license to purchase or use the equipment, there are usually regulations in place that prevent the sale radios if they have not undergone testing by the licensing body [48]. This type of rule helps to ensure that users do not accidentally operate outside of the restrictions placed on the license exempt band.

Bands that are "licensed by rule" operate in much the same that license exempt bands do but also require the user to obtain a blanket license that covers all operations within a shared band. Transmissions within these bands are not specifically assigned by the regulatory bodies, the entire band is typically open to any licensed user provided they follow the rules and restrictions of their particular certification. For example, the maritime and amateur bands in the US are free and open to use by anyone who has passed the necessary examination requirements. These licenses, usually awarded to anyone who can pass the required examination, are primarily used to ensure that users of the shared bands are aware of the rules and regulations regarding the type and content they are allowed to transmit.

2.3.3 Industrial, Scientific, and Medical Band

Though wireless technologies were originally envisioned solely as communications technologies, primarily as a method for replacing telegraph lines, research quickly showed there could be a number of uses for electromagnetic waves in other scientific disciplines. For example, wireless technologies emerged that could be used for remotely detecting objects, heating up food, or wirelessly transmitting electrical power. It stands to reason that if governments are going to regulate the transmission of wireless signals for communication that they must similarly regulate the emissions from noncommunications emitters or the benefit of regulation is lost. If communication bands are neatly partitioned, but users of microwave ovens and RADAR transmitters are allowed to operate freely then there will inevitably be significant interference between any communications and sensing equipment operating in the same band. To this end, the US proposed the international allocation of certain spectrum bands for "Industrial, Scientific, and Medical applications" at the 1947 ITU Radio Conference in Atlantic City, NJ, USA [2]. It was at this conference that the first ISM bands,were allocated for "A service other than a radiocommunication service, for industrial, scientific, or medical uses, which results in the transmission of energy by radio." [49] On the recommendations of research performed by US and French scientists, the ITU reserved four bands for ISM usage at this conference in the 13, 27, 41 MHz, and 2.4 GHz ranges. It is important to note that the ITU did not specifically outlaw the use of the radios in these bands, but stipulated that "Radiocommunication services operating within those limits must accept any harmful interference that may be experienced from the operation of industrial, scientific and medical equipment." This meant that member countries were still free to assign secondary allocations as they saw fit, but that there would be no guarantees that microwave ovens from across the border would not interfere with their communications.

Because the ITU language for these ISM bands was very specific, the only real restriction on devices operating in these bands was that it must accept interference from other sources within the band. Given the original justification for the ISM band, namely remote heating and power transfer applications, **accepting interference** would be considered a *given* in the sense that any power in the band only enhances these use cases rather than leading to a degradation in effectiveness. This is in direct opposition to communication technologies where additional energy in the band is almost always considered harmful and interferes with the intended use. However, much like maritime radar and other *geographically sparse* services, the intended applications for ISM devices leaves a wide swath of potentially usable spectrum sitting empty over large geographical areas. In addition, the lack of licensing requirements provide a low barrier of entry to adoption for technologies that do not require guaranteed protections. For this reason, the ISM bands have rapidly become home for a wide array of short range wireless communication technologies that lack strict quality of service requirements. These use cases include wireless access control mechanisms that replace physical keys, short-range wireless networking for computers, wireless interconnection of personal peripherals such as mice keyboards and headphones, home automation and control, and many more. The most successful use case is undoubtedly short-range wireless networking using the IEEE 802 family of standards such as 802.11, 802.15.1, and 802.15.4. Operating almost exclusively in the ISM bands, these technologies have exploded due to their low barrier to entry and open licensing schemes. By allowing users to simply "plug and play" without having to concern themselves with obtaining a license, these short-range wireless services have exploded in popularity for home and office use. The result is that these unlicensed bands, originally set aside as a *no man's land* where communications would not be possible due to interference, have become one of the most used communications bands in the world.

The massive rise in popularity of these short-range networking services that operate in the ISM has led to massive congestion in these bands and resulted in the allocation of new bands specifically designed for use with unlicensed network devices [50]. This rise in popularity also comes with a downside. As the adoption of these technologies increases, countries are left with the traditional *tragedy of the commons* problem that radio regulations sought to eliminate. Because these devices are not centrally managed, they are forced to use inefficient schemes for accessing the spectrum and coordinating transmissions between nodes. This means that the number of networks that can operate at any given time is lower than could be achieved via central control. While there are many techniques to mitigate this congestion, the fact remains that these technologies must ultimately rely on a naive *listen before talk* media access scheme due to the fact that interference from ISM devices may come at any time. This fundamental fact [51] is what drives research into spectrum sharing and seeks to find a better way to share unlicensed spectrum between multiple users.

2.3.4 TV White space

At the start of the twenty-first century, governments around the world began the process of transitioning broadcast television services from traditional analog video to newer digital standards. This switch allowed for a much more efficient use of the spectrum when transmitting the same number of television channels and ultimately led to what the ITU refers to as the *Digital Dividend*. The **Digital Dividend** refers to the contiguous chunks of spectrum freed up during the consolidation of the old channel band plans. During the period leading up to the ITU's re-allocation of these bands, there was significant discourse around the best way to recapture this valuable but unused spectrum in the UHF band [52]. There were several proposals that afforded differing degrees of backwards compatibility and international harmonization of frequencies. The first proposal, ultimately adopted, includes re-combining all television broadcast channels into the lower half of the band and freeing the upper half of the band for use in traditional, cellular, mobile telephony allocations. There was another proposal, however, that sought instead to leave the bands as allocated but with the addition of a secondary service that would allow the ad-hoc use of the unoccupied television channels. This second proposal was attractive because it freed up the spectrum for use while maintaining 100% backwards compatibility with existing infrastructure meaning there would be no costs to replace transmission equipment or obsolete televisions. The primary drawback of this scheme is that it would be impossible for member countries to allocate the unused spectrum in logical chunks for traditional mobile telephony paradigms. New standards and equipment would have to be developed that allowed the ad-hoc use of this potentially free spectrum without interfering with local broadcast television stations. Ultimately, the band consolidation took place, resulting in two separate changes. The first, at World Radio Conference 2007 (WRC-07), freed the blocks ranging from 790 to 862 MHz in Region 1 and 698–806 MHz in Region 2 to be used for mobile telephony services [53]. The second round, ratified at the World Radio Conference 2012 (WRC-12) freed an additional band for Region 2 in the 694–790 MHz spectrum.

Even though the concept of TV white space technology did not serve as sufficient motivation to prevent re-allocation in the television bands, there still remains a substantial amount of unused spectrum in the newly condensed digital television bands at 470–694 MHz. This underutilized spectrum exists for all the same reasons that vacant channels existed in the analog TV broadcast band, namely frequency separation between neighboring broadcast cells and a lack of broadcast channels in a given geographic area. The same concepts for sharing the TV white space that were proposed for opportunistic use in the analog allocations can be applied in the new digital broadcast bands, albeit with less overall economic value [54]. This point is further driven home by the fact that during the same WRC-12 meeting, studies were presented that determined the current ITU Radio Regulations are sufficiently flexible to allow for the white space radios to operate within the recently reduced television bands [55]. This means that there is a real possibility for countries to adopt the opportunistic use of the underutilized digital television spectrum as they see fit while maintaining interoperability with the rest of the world.

2.3.5 CBRS

As part of the NTIA's 10 year plan for free up US spectrum, a particular 150 MHz wide section, ranging from 3500 to 3650 MHz, was identified as a candidate to be *fast tracked* for transition from exclusive federal allocations to the commercial and civilian use [56]. As of 2015 this section of the spectrum is now called the **Citizens Broadband Radio Service Band** (CBRS band) and allows for a variety of uses simultaneously benefit the incumbents, commercial use, and public access. The motivation for fast tracking this band centered primarily on the usage models of the incumbents which were largely separated into two groups consisting of Naval ocean radars and fixed satellite terminals. In both cases, the locations of these incumbents is very predictable within the regulatory boundaries of the United States. First, the high powered naval radars using this band are, by nature, going to be limited to portions of the country that are close to the ocean. This leaves a large sectional area of the in-land US territory where these frequencies would otherwise be wasted. This, in fact, is the reason that the second incumbent group—Fixed Satellite Services—are able to effectively share the band with the naval radar users. Because they are in-land there is no interference. However, because these frequencies are generally used by large data providers and require high gain antennas, there only are few sites distributed around the US. According to the FCC's database, in January 2020 there were only 286 active sites. In addition, because these services are often managed by the same coordinators they can often be co-located. Using the same data, 155 of the active antennas were distributed across only four distinct regional sites [57]. What this tells us is that even though these incumbent users are easily interfered with, they are few and far between with respect to the vast geographic space in the USA. This relatively narrow distribution of incumbent users combined with the low probability of spectrum usage by naval radars within the US interior were the primary reasons this band was chosen as a candidate for fast track [56].

First released in 2015, Title 47 Part 96 of the of the US Code of Federal Regulations outlines the rules and regulations required to operate within the CBRS bands inside the US [58]. Under part 96, the incumbent access to the spectrum is ensured via a frequency coordination system called the Spectrum Access System (SAS) that utilizes a combination of a location database and environmental sensors that estimate incumbent usage at fixed points. In addition to incumbent users, part 96 introduces two additional tiers of access that can be used for commercial and public uses. This band, then, can be used by three classes of users, called **tiers**. These tiers are delineated by their priority when accessing the spectrum: *Incumbent Access Users*, *Priority Access Users*, and *General Authorized Access Users*.

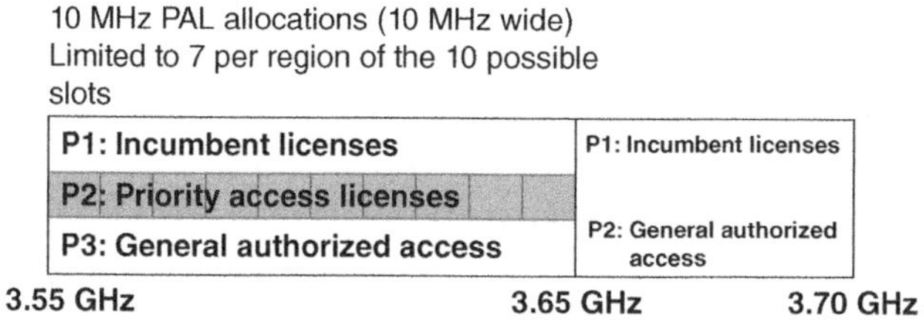

10 MHz PAL allocations (10 MHz wide)
Limited to 7 per region of the 10 possible
slots

P1: Incumbent licenses	P1: Incumbent licenses
P2: Priority access licenses	
P3: General authorized access	P2: General authorized access

3.55 GHz　　　　　　　　　　3.65 GHz　　　　　3.70 GHz

Figure 2.6 Band plan for CBRS.

The first tier, **Incumbent Access Users**, are the naval radars and existing satellite services that register with the SAS to ensure no licenses are granted to the other tiers that might interfere with their operations. The second tier, **Priority Access**, provides Priority Access Licenses (PAL) that grants exclusive access to one 10 MHz chunk of the band in one of the 3234 counties under US control [59]. These 10 MHz wide bands allow for 10 potential slots across the allowed band as seen in Figure 2.6. However, in order to promote open and fair access, the FCC restricts the number of total licenses granted in any given county to seven and the number of licenses to a single user to four. These caps ensure there is both spectrum available for public use everywhere and that no one company may hold a monopoly on the valuable spectrum in a given area. The third tier, **General Authorized Access** provides citizens with additional spectrum that will provide the same benefits as have historically been provided by the ISM band. This tier is treated the same as users of the ISM band in the sense that they must accept any interference on the band with no expectation of guaranteed access. However, devices certified for use in this band have the additional requirement that they guarantee not to cause harmful interference to the incumbent and priority tiers operating in the same band. Based on the FCC rules for granting PALs, this means that there will be a minimum of 30 MHz of bandwidth across the US that will be available for general access with the potential for more in areas that do not have the maximum of seven PALs.

Overall, the part 96 rules for the CBRS achieve the goals set out by the NTIA and President's Council of Advisors on Science and Technology (PCAST) in the sense that a large section of useful spectrum was freed for both commercial and public use while maintaining the interests of incumbents. In addition, the three service tiers seem to adequately balance the interests of the federal government in the sense that revenue is still generated through licensing auctions while public access to the shared spectrum resources is still available. With the first PALs set to be auctioned in July of 2020 [60] and initial GAA devices hitting the market, CBRS will be an important testing ground for examining the best way to balance the competing interests of commercial and public spectrum allocations.

References

1 ITU Telecommunication Development Sector (2012). *Exploring the Value and Economic Valuation of Spectrum*. International Telecommunications Union.

2 (1947). Radio regulations and additional radio regulations. In: *International Radio Conference*. Atlantic City, NJ: The International Telecommunication Convention.

3 United States. Dept. of Commerce. Bureau of Navigation.(1914). Radio communication laws of the United States and the international radiotelegraph convention: regulations governing radio operators and the use of radio apparatus on ships and on land. In:*Conference on Wireless Telegraph (1912) London, E., United States*.

4 Radio Chaos ends tomorrow night. *Evening Star,* p. 2, 22 April 1927. https://chroniclingamerica.loc.gov/lccn/sn83045462/1927-04-22/ed-1/seq-2/.

5 Davis, S. (1927). *The Law of Radio Communication*. New York: McGRaw-Hill.

6 Bureau of Navigation, Department of Commerce USA (1915). *Radio Service Bulletin*. Washington, DC: The Bureau.

7 Bensman, M.R. (2000). *The Beginning of Broadcast Regulation in the Twentieth Century*, 1e. McFarland & Company. (1 March 2000).

8 Titrle 47 section 153. [Online]. https://uscode.house.gov/view.xhtml?req= (title:47%20section:153%20edition:prelim (accessed 20 September 2020).

9 Basic elements of spectrum management. [Online]. https://www.ntia.doc.gov/legacy/osmhome/roosa4.html (accessed 7 October 2019).

10 A short history of NTIA. [Online]. https://www.ntia.doc.gov/legacy/opadhome/history.html (accessed 7 October 2019).

11 US Reports (1945). *Ashbacker Radio Co. v. FCC*. Washington, DC.

12 Blake, J. (1994). FCC licensing: from comparative hearings to. *The Federal Communications Law Journal* 47 (2): 7. https://www.repository.law.indiana.edu/fclj/vol47/iss2/7/.

13 Federal Communications Commission. Auction 73: 700 MHz band. [Online]. https://www.fcc.gov/auction/73/factsheet (accessed 7 October 2019).

14 (1927). *International Radiotelegraph Conference*. Washington, DC: The International Telecommunication Convention http://handle.itu.int/11.1004/020.1000/4.39.

15 (1912). *The Final Protocol*. London: The International Telecommunication Convention http://handle.itu.int/11.1004/020.1000/4.37.

16 L.S. Howeth.*History of Communications-Electronics in the United States Navy*. University of Michigan Library (1 January 1963).

17 Complete list of radio conferences. [Online]. https://www.itu.int/en/history/Pages/CompleteListOfRadioConferences.aspx (accessed 7 October 2019).

18 Abramson, N. (1985). Development of the ALOHANET. *IEEE Transactions on Information Theory* 31 (2): 119–123. https://doi.org/10.1109/TIT.1985.1057021.

19 TIA/EIA/IS-95-A (1995). *Mobile Station – Base Station Compatibility Standard for Dual-Mode Wideband Spread Spectrum Cellular System*. Telecommunications Industry Association.

20 Tanenbaum, S. (2003). *Computer Networks*. Upper Saddle River: Prentice Hall.

21 IEEE 802 (1999). *IEEE Std 802.11b*. IEEE.

22 Penny Pritzker, S. (2016). *Quantitative Assessments of Spectrum Usage*. Washington, DC: U.S. Department of Commerce.

23 S.L. Herzel. (1951).*Public Interest and the Market in Color Television Regulation*. 8 U. Chi. L. Rev.802.

24 US SUpreme Court.Dolan v. City of Tigard 512 U.S. 374, 384.

25 G.G. Stevenson. *Common property economics: a general theory and land use applications*. Cambridge University Press, 1991. 0 521 38441 9. *Ecological Economics*, Elsevier, vol. 9(2), pages 181–182.

26 The Law and Economics of Property Rights to Radio Spectrum (1998). Diminnative Creation of Property Rights to the Radio Spectrum. *The Journal of Law and Economics* 41 (S2) https://doi.org/10.1086/467404.

27 Robinson, G.O. (1998). Spectrum property law 101. *The Journal of Law and Economics* 41 (S2): 609–626.

28 P.R. Milgrom, J.D. Levin, and A. Eilat. (2011).*The Case for Unlicensed Spectrum*. SSRN. https://ssrn.com/abstract=1948257 (accessed 7 October 2019).

29 Hardin, G. (1968). The tragedy of the commons. *American Association for the Advancement of Science* 162 (3859): 1243–1248.

30 (1903). *Preliminary Conference on Wireless Telegraphy*. Berlin: The International Telecommunication Convention https://doi.org/10.1004/020.1000/4.35.

31 ITU membership overview. [Online]. https://www.itu.int/en/membership/Pages/overview.aspx. (accessed February2020).

32 International Telecommunications Union.(2012). *ITU Radio Regulations Articles*.

33 World radiocommunication conferences (WRC). [Online]. https://www.itu.int/en/ITU-R/conferences/wrc/Pages/default.aspx. (accessed February2020).

34 About CITEL. [Online]. https://www.citel.oas.org/en/Pages/About-Citel.aspx. (accessed February2020).

35 CEPT electronic communications committee. [Online]. https://www.cept.org/ecc. (accessed February2020).

36 A new dimension of challenge for Asia-Pacific telecommunity. [Online]. https://www.apt.int/APT-Introduction. (accessed February2020).

37 History of the African telecommunications union. [Online]. http://atu-uat.org/history. (accessed February2020).

38 (1995). *S-A.1154: Provisions to Protect the Space Research (SR), Space Operations (SO) and Earth Exploration-Satellite Services (EESS) and to Facilitate Sharing with the Mobile Service in the 2 025–2 110 MHz and 2 200–2 290 MHz Bands*.

The International Telecommunication Union https://www.itu.int/dms_pubrec/itu-r/rec/sa/R-REC-SA.1154-0-199510-I!!PDF-E.pdf.

39 NTIA. (2014).*NTIA Alloctation for 2200–2290 MHz.*

40 Hart, J. (2011). *The Transition to Digital Television in the United States: The Endgame.* International Journal of Digital Television.

41 Ludwig, R. and Taylor, J. (2002). *Voyager Telecommunications.* Pasadena, CA: NASA JPL DESCANSO.

42 (1995). *SA.1154: Provisions to Protect The Space Research (SR), Space Operations (SO) and Earth Exploration-Satellite Services (EESS) and to Facilitate Sharing with the Mobile Service in the 2 025–2 110 MHz and 2 200–2 290 MHz Bands.* The International Telecommunication Union https://www.itu.int/dms_pubrec/itu-r/rec/sa/R-REC-SA.1154-0-199510-I!!PDF-E.pdf.

43 (2014). *Fourth Interim Progress Report on the Ten-Year Plan and Timetable and Plan for Quantitative Assessmentsof Spectrum Usage – U.S. Department of Commerce.* The National Telecommunications and Information Administration https://www.ntia.doc.gov/report/2014/fourth-interim-progress-report-ten-year-plan-and-timetable.

44 (2015). *Guidelines for the Preparation of a National Table Offrequency Allocations (NTFA). ITU Telecommunication Development Sector.* The International Telecommunication Union.

45 Restrepo, J. (2016). *Radio Regulations.* Geneva: World Radiocommunication Seminar 2016.

46 ITU-D Study Group 1, "*Evolving Spectrum Management Tools to Support Development Needs,*" 2017.

47 *Case Study: FCC License-by-Rule.*Administrative Conference of the United States. https://www.acus.gov/sites/default/files/documents/Licensing%20and%20Permitting%20Appendix%20B.pdf.

48 (2020). *Code of Federal Regulations Title 47: Telecommunication Part 2—Frequency Allocations and Radio Treaty Matters; General Rulesand Regulations Subpart J—Equipment Authorization Procedures.* https://blog.apastyle.org/apastyle/2013/07/the-rules-for-federal-regulations-i-code-of-federal-regulations.html.

49 (1947). *Documents of the International Radio Conference (Atlantic City, 1947).* Atlantic City, NJ: The International Telecommunication Union http://handle.itu.int/11.1004/020.1000/1.7.

50 5 GHz unlicensed spectrum (UNII). (2013). [Online]. https://www.fcc.gov/document/5-ghz-unlicensed-spectrum-unii. (accessed February2020).

51 Kleinrock, R.L.a.L. (2016). The capacity of wireless CSMA/CA networks. *IEEE/ACM Transactions on Networking* 24, 3: 1518–1532.

52 (2012). ITU telecommunication development sector. *Digital Dividend: Insights for Spectrum Decisions.* https://www.itu.int/en/ITU-D/Technology/Documents/Broadcasting/DigitalDividend.pdf

53 Kholod, A. (2010). *The Digital Dividend Opportunities and Challenges*. International Telecommunications Union.

54 (2018). *Digital Dividend: Insights for Spectrum Decisions. International Telecommunications Union*. The International Telecommunication Union https://www.itu.int/en/ITU-D/Spectrum-Broadcasting/Documents/Publications/DigitalDividend_Final_2018.pdf.

55 Al-rashedi, N. (2014). *Spectrum Management Studies, Incl. RA-12 & WRC-12 Related Outcomes*. Geneva: The International Telecommunication Union https://www.itu.int/en/ITU-R/study-groups/workshops/RWP1B-SMWSCRS-14/Presentations/SG%201%20-%20Spectrum%20management%20studies,%20incl.%20RA-12%20and%20WRC-12%20related%20outcomes.pdf.

56 US Department of Commerce. (2010)*An Assessment of the Near-Term Viability of Accommodating Wireless Broadband Systems in the 1675–1710 MHz, 1755–1780 MHz, 3500–3650 MHz, and 4200–4220 MHz, 4380–4400 MHz Bands*.

57 FCC open data – protected FSS earth station registration. [Online]. https://opendata.fcc.gov/Wireless/Protected-FSS-Earth-Station-Registration-Complete-/acbv-jbb4/data (accessed 7 October 2019).

58 Code of Federal Regulations Title 47 → Chapter I → Subchapter D → Part 96 Title 47: Telecommunication PART 96—Citizens Broadband Radio Service.May 2020. [Online]. https://www.ecfr.gov/cgi-bin/text-idx?node=pt47.5.96&rgn=div5 (accessed 7 October 2019).

59 2017 counties (2020 05 14). [Online]. https://www.fcc.gov/file/18824 (accessed 7 October 2019).

60 Auction 105: 3.5 GHz. [Online]. https://www.fcc.gov/auction/105/factsheet (accessed 7 October 2019).

3

Concepts in Communications Theory

Communications theory, as a whole, is a broad subject. It involves radio hardware, such as antennas and RF circuits. It involves a physical and media access layer concepts such as propagation and modulation. It extends to networking concepts.

These concepts in communications theory can be esoteric. For the uninitiated reader, some of the concepts necessary to understand the discussion on wireless coexistence may be unapproachable without some primer. It is the intention of this chapter to provide that primer. This chapter will present concepts in communications theory specifically relevant to the discussion on wireless coexistence within this book. The topics selected to be covered in this chapter will be concepts revisited in later chapters.

To have any fruitful discussion, a common set of definitions must be used. Wherever possible, this book will defer to the definitions established in the standard IEEE 1900.1 [1]. This reference does not provide consistent definitions for all concepts to be discussed in this book. This chapter will provide definitions for commonly used terminology, such as different types of *channels.*

Some of the standards discussed in this book define an air interface. Many of those air interfaces employ a multicarrier modulation scheme called orthogonal frequency division multiplexing (OFDM). This chapter will provide an overview of that multicarrier modulation scheme and the linear modulation on the individual subcarriers.

In addition to detailing the selected topics, this chapter will provide a list of references for the interested reader such that they may delve deeper into a given subject.

Wireless Coexistence: Standards, Challenges, and Intelligent Solutions, First Edition.
Daniel Chew, Andrew L. Adams, and Jason Uher.
© 2021 The Institute of Electrical and Electronics Engineers, Inc.
Published 2021 by John Wiley & Sons, Inc.

3.1 Types of Channels and Related Terminology

IEEE 1900.1 does not provide a consistent definition of *channel*. The word "channel" gets used in almost all of communications theory literature but the exact meaning is entirely context-dependent. Often the word "channel" is decorated with a qualifier such as "frequency" to become *frequency channel* or "logical" to become *logical channel*. The definition of such qualifiers also depends on context or a putative meaning. If the reader is not familiar with a term, such reliance on putative meanings is not helpful.

This section will provide definitions for various types of channels. These terms will be used throughout the book.

A **frequency channel** refers to a specific allocation in the spectrum. It consists of a center frequency and bandwidth. A frequency channel is sometimes called a *broadcast channel*. The term *spectrum channel* means frequency channel in TV Whitespace (TVWS) literature. Concurrent users in a band can be separated using different center frequencies and mutually exclusive bandwidths, i.e., each user gets a frequency channel on which to transmit. Receivers can tune to one of the frequency channels and receive only data from that source. In this way, a frequency channel can be both a physical channel (center frequency and bandwidth) and a logical channel (Affiliated with a source of data). Historically, primary users were those to whom a regulatory authority gave exclusive permission to use some frequency channel at some location. Broadcast radio and television stations are a classic example of this phenomenon. Frequency channels may be specified and numbered in a given band. An example of such frequency channels would be the IEEE 802.11 channels [2].

A **logical channel** refers to desired user data selected out of a set of signals or a composite signal. If a single frequency channel contained data for 10 different users, each of the 10 users would be assigned a logical channel within that frequency channel. Receivers of that signal would differentiate between the data intended for each individual user. Some mechanism needs to be employed to differentiate the users sharing a frequency channel. Time slots are one such mechanism, where every user gets a pre-allocated amount of time on the shared frequency channel. Orthogonal codes can be used to separate users sharing a frequency channel, and in doing so form logical channels.

A **physical channel** refers to a channel that has a physical definition. The physical channel represents the physical medium over which the signal travels.

A **control channel** is either a frequency channel or a logical channel on which a network coordinator sends control information to connected nodes.

A **hop channel** is a specific frequency to which a frequency hopping system has hopped. A hop channel is generally part of a *hop set*, that being a set of frequencies over which the system will traverse.

3.2 Types of Interference and Related Terminology

The terms **noise** and **interference** refer to undesirable signals impeding the ability of a receiver to recover the desired signal. Interference is assumed to be man-made, whereas noise is a natural phenomenon. For example, it may be said that two transmitted channels *interfere* with one another.

Co-channel interference (CCI) occurs when two signals occupy the same frequency channel.

Adjacent channel interference (ACI) occurs when two signals occupying adjacent frequency channels interfere with one another.

Intersymbol interference (ISI) occurs when two symbols from the same transmitter interfere with one another in time. This can occur in multipath channels where a symbol is *smeared* past the intended period of that symbol and into the period of another time-adjacent symbol.

Intercarrier interference (ICI) occurs when two adjacent subcarriers in a multicarrier modulation scheme interfere with one another. The multicarrier modulation scheme relies on the *orthogonality* of the individual subcarriers. There are impairments which cause a loss of this orthogonality, and adjacent subcarriers can then interfere with one another.

3.3 Types of Networks and Related Terminology

One of the first things that is needed when defining a standard for a wireless system is to determine over how wide an area that system will provide data connectivity.

The **spatial scope** of a network determines how wide an area is to be covered by that network.

Personal area networks (PANs) are networks very limited in spatial scope and may only cover devices around a single user. An example of a wireless PAN (WPAN) would be the connection between a smart phone and a Bluetooth headset. Coexistence standards that are found in modern Bluetooth, such as Adaptive Frequency Hopping from IEEE 802.15.2 [3], will be discussed in Chapter 8.

Local area networks (LANs) are networks that are intended to provide connectivity to the home or office. Indoor propagation would be important in the analysis of a wireless LAN (WLAN). An example of a WLAN would be the IEEE 802.11 standard [2], which will be discussed in Chapter 8.

Regional area networks (RANs) are intended to range over many kilometers. Propagation effects like shadowing are important in the analysis of a wireless RAN

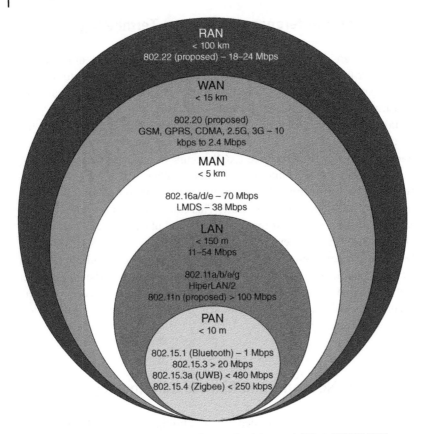

Figure 3.1 Network types and sizes. *Source:* Cordeiro et al. [5]. © [2020] IEEE.

(WRAN). An example of a WRAN would be the IEEE 802.22 standard [4], which will be discussed in Chapter 8.

An illustrated comparison of network types based on spatial scope comes from Reference [5] and is shown in Figure 3.1. This figure illustrates the various types of networks based on the range of coverage. The comparison begins with personal area networks (PANs) as the smallest and builds up to regional area networks (RANs). When these networks are wireless, a W is placed before the acronym such as WPAN and WRAN.

These networks often work in concert with one another; the idea being that a local area network (LAN) in one's home or office provides that user with connectivity to a RAN. Reference [5] was written in 2006 and therefore some of the comments in the figure are out of date, such as 802.11n being "proposed."

3.4 Primer on Noise

The additive white Gaussian noise (AWGN) channel, and the physical phenomenon being modeled by it, is discussed in a great many texts on communications theory. Recommendations for a reader needing a deeper understanding include References [6] and [7]. This section will provide a brief primer to familiarize the reader with the concept of AWGN so as to provide a background for more complex topics discussed in this chapter.

The model of noise within the bandwidth of a receiver is illustrated in Figure 3.2 where a noise signal, $n(t)$, is added to the received signal. The noise in the receiver bandwidth is referred to as AWGN. This is because this AWGN noise is:

- Added to the received signal,
- It is flat across all frequencies,
- And it is Gaussian in its probability distribution.

The model illustrated in Figure 3.2 shows that AWGN is a phenomenon that occurs inside the receiver. AWGN is not affected by multipath or propagation. AWGN is sometimes called *thermal noise* as the source of the noise is a product of the receiver electronics and the power increases with temperature.

Consider a line-of-sight wireless link, meaning there are no obstructions between the transmitter and the receiver. As the signal propagates, the signal loses power. Therefore, the signal as received at the receiver will be much lower in power than it was when transmitted at the transmitter. This scenario poses the question: is there any lower limit to the received power in order for the receiver to reliably demodulate the signal? The fundamental factor limiting the ability of a receiver to successfully demodulate a signal is noise within the receiver bandwidth. The noise power in the receiver bandwidth will dictate the power of the *minimum detectable signal* (MDS).

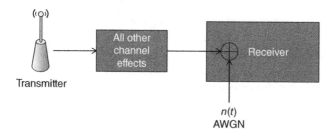

Figure 3.2 The AWGN channel in a communication system.

3.5 Primer on Propagation

The standards discussed in this book involve both long-range and short-range communication systems. For example, IEEE 802.11 [2] is a wireless local area network (WLAN) originally intended to provide wireless connectivity in the home or office, whereas IEEE 802.22 is a wireless regional area network (WRAN) intended to provide wireless connectivity over a range of many kilometers. These two waveforms are intended for different environmental scenarios and thus are designed to overcome different impairments. There are many great references on wireless propagation. It is not the intention of this section to duplicate that body of work. Instead, this section will provide a brief primer on wireless propagation highlighting concepts that are relevant to this book's discussion on wireless coexistence. This section will identify channel models used for relevant standards and provide a list of references for readers who want to delve more deeply into the topic of wireless propagation.

A wireless signal must propagate from the transmitter to the receiver. Along the way, the transmitted signal will lose power and be subject to *channel impairments*. These channel impairments include *large-scale fading* and *small-scale fading*. **Large-scale fading** models the loss of signal power over large distances between the transmitter and the receiver. **Small-scale fading** represents the changes in the received power that are resulting from small changes in the general positioning between a receiver and a transmitter. Thus, the terms "large" and "small" refer to the magnitude of the changes in distance between the receiver and the transmitter. To put this into context, large-scale fading would model the general loss in a Wi-Fi signal measured some radius away from that transmitter. Small-scale fading would model the differences between measures given small changes in distance for receivers placed near that established radius.

All the waveform standards addressed in this book specify a minimum receiver power and define a maximum BER allowable at that minimum power. For the purpose of discussing link budget analysis, the signal at that minimum power will be referred to as the *minimum detectable signal*. A **link budget** determines the received power of a wireless link over some range taking into account the transmit power, antenna gains, propagation losses, and a margin for channel impairments. A link budget can be used to determine the expected received power given some distance and some physical scenario (Urban, Rural, etc....). The link budget can help determine the maximum distance achievable between the transmitter and the receiver to maintain the minimum detectable signal given a model of the wireless environment.

An excellent description these concepts can be found in Reference [8]. Figure 3.3, from Reference [8], illustrates each contributing factor. Distance between the

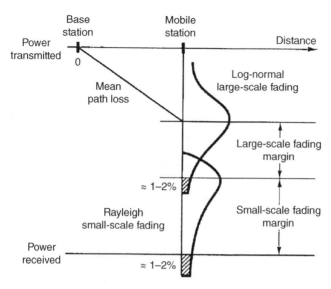

Figure 3.3 Link budget analysis. *Source:* Sklar [8]. © [2020] IEEE.

transmitter and the receiver is shown in the *x*-axis. Loss in the signal power is shown in the *y*-axis. The first effect illustrated in Figure 3.3 is labeled *mean path-loss* and that is the power lost due to propagation, and that is deterministic.

The path-loss can be computed as follows:

$$L_p = n*10 \log_{10}(d) + 20 \log_{10}\left(\frac{4\pi}{\lambda}\right) \tag{3.1}$$

where

λ is the wavelength of the wireless signal,
d is the distance between the transmitter, and
n is the large-scale fading exponent

The wavelength is a factor in path-loss because the wavelength determines the size of the antenna. Antennas are designed to accommodate the intended wavelength. Wavelength and frequency are inversely proportional; therefore, a larger wavelength means a lower frequency. Larger antennas provide a larger aperture, thus more area to illuminate.

The *mean path-loss* in Figure 3.3 provides a mean for the nondeterministic large-scale fading (*shadowing*). Shadowing is a type of channel impairment that represents random obstructions along the path of the transmission. Shadowing is modeled as a log-normal random loss centered at the expected power level left after deterministic path-loss.

Small-scale fading does not mean small changes in the received power but rather refers to small changes in the distance or positioning between the receiver and the transmitter. Small-scale fading accounts for multipath effects. A multipath environment is one where the transmitted signal is reflected and/or scattered by objects along the path of the transmission. The result is that multiple paths of the transmitted signal add at the receiver. These paths therefore interfere with one another at the receiver. Each path represents the transmitted signal with a different delay. Due to these different delays, the paths may sum together constructively or destructively (constructive and destructive interference). Consider signals in the 2.4 GHz ISM band. A signal with a center frequency of 2.4 GHz has a wavelength of 0.125 m. If one path is 0.0625 m longer than another path, then the two will be 180° out of phase and may cancel each other out completely. Therefore, small changes in distance may have dramatic effects on the received signal power. Small-scale fading is modeled as stochastic (random) process. In a link budget analysis, as shown in Figure 3.3, loss in signal power due to small-scale fading is added to the loss due to large-scale fading.

The goal of the link budget analysis is to establish a **link margin**, defined as the amount of power in the received signal above the minimum-detectable signal, given distance, and environmental parameters. In Figure 3.3, the link margin is intended to provide adequate received signal power, that being at or above the minimum-detectable signal, for the vast majority of cases. This requires planning for cases that are near to the worst case. For example, Figure 3.3 plans for 98–99% of all cases of hypothetical environment. To meet these criteria, the mean of the probability distribution of the small-scale fading is placed at the 98% mark of the large-scale fading. The expected received power is then taken from the 98–99% mark of the small-scale loss. This planning results in an adequate margin in 98–99% of all possible scenarios.

Multipath channel models for the waveforms discussed in this book have been researched and provided by the standards bodies. One of the earliest relevant to the waveforms here is from IEEE 802.11 and illustrated in Figure 3.4 from Reference [9]. The figure details a *power-delay profile*. The power-delay profile can be considered coefficients in finite impulse response (FIR) filter. Thus, the channel has a *channel impulse response*. In this example, the channel is composed of multiple complex-valued components where the magnitude is Rayleigh distributed and the phase is uniformly distributed. The black lines represent the expected magnitude and delay. The gray lines indicate a specific instance of the random channel. The Rayleigh distribution is often used as a representative of the *worst-case scenario* in wireless communications because it very nicely models a situation in which there is no direct line of sight between the transmitter and the receiver, and all of the received energy is reflected off of different objects in the environment.

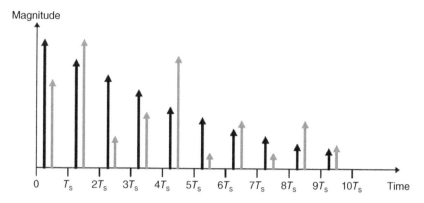

Figure 3.4 Power-delay profile for 802.11. *Source:* Based on Chayat [9].

The channel model for the IEEE 802.11 standard continued to evolve over time resulting in a far more complex system of delay profiles enumerated in Reference [10] as models A through F. The IEEE 802.11 channel model was further developed in task group ac (TGac) in Reference [11]. The channel model used by task group af, which dealt with the use of IEEE 802.11 in TV Whitespace, borrowed heavily from IEEE 802.22 as discussed in Reference [12]. The model for IEEE 802.22 is defined in Reference [13].

Another important point about multipath channels is that they exhibit *reciprocity*. A given wireless channel only exists between two nodes. This is because the two nodes have a unique spatial position.

A multipath channel may be **frequency selective**, meaning that the frequency response of the channel significantly attenuates spectral components of the signal passing through it. Thus, the determination of whether or not a channel is frequency selective is relative to the bandwidth of the signal one intends to pass through the channel and relative to the frequency-channel being used. The phenomenon is illustrated in Figure 3.5 from Reference [8]. The variable W is the bandwidth of the signal to pass through the channel. The variable f_0 in the figure is called the **coherence bandwidth** of the channel. Coherence bandwidth is a statistical measure designed to indicate a bandwidth over which the channel is not likely to distort the spectral components of the signal. Coherence bandwidth is inversely proportional to the maximum delay in the power delay profile. That is to say that the power delay profile represents the impulse response of the channel, and the coherence bandwidth is a measurement made from the frequency response of the channel. Therefore, the greater the delay between two significant rays in the power delay profile, the more narrow the bandwidth exhibited by the frequency response of that channel.

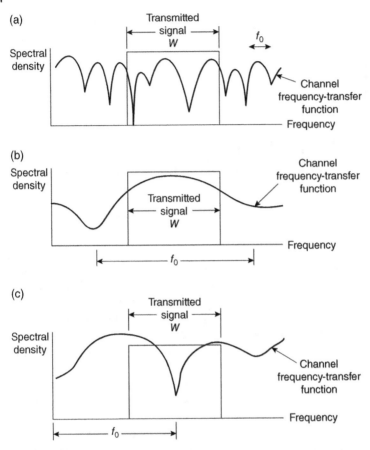

Figure 3.5 Frequency selectivity [8]. (a) Typical frequency selective fading case ($f_0 < W$). (b) Typical flat-fading case ($f_0 > W$). (c) Null of channel frequency-transfer function occurs at signal band center ($f_0 > W$). *Source:* Sklar [8]. © [2020] IEEE.

Figure 3.5 shows examples of different channel types. The first diagram shows a *frequency selective* channel which is defined as the situation where the coherence bandwidth of the channel is less than the bandwidth of the signal, or $f_0 < W$. The second diagram in Figure 3.5 shows a *flat-fading* case, which is when the channel does not significantly attenuate or distort any of the spectral components of the signal. This means the coherence bandwidth of the channel is greater than the bandwidth of the signal, or $f_0 > W$. However, it is not always the case that a coherence bandwidth greater than the signal bandwidth will yield a flat fading channel. The last diagram in Figure 3.5 shows a situation where the coherence bandwidth of the channel is greater than the signal bandwidth, but the channel still exhibits frequency selectivity. In the last case, the channel has large stretches of spectrum without frequency selectivity. However, even though the average is flat, there is a null in the frequency response, which interferes with the signal of interest.

Immediately from the equations $f_0 < W$ and $f_0 > W$, one can see that a signal with a lower symbol rate (less bandwidth) will have an easier time in avoiding frequency selectivity. It is noted that this statement refers to the symbol rate and infers a larger bandwidth. As will be seen in the discussion on spread spectrum, it is sometimes advantageous to increase the bandwidth of the signal to greater than that required by the symbol rate. Doing so provides *frequency diversity*, which grants resilience to frequency selectivity.

3.6 Primer on Orthogonal Frequency Division Multiplexing

Many of the standards covered in this book employ orthogonal frequency division multiplexing (OFDM). Therefore, some discussion of this technique is important for the reader who may not be familiar with the concept. This section will provide a brief primer on OFDM and provide further references for a reader interested in diving deeper into the topic.

OFDM is a multicarrier modulation technique in which the transmitter linearly modulates multiple *orthogonal* subcarriers concurrently. The process begins with mapping data bits onto *symbols* which are *complex-valued*, then using those symbols to modulate a set of *orthogonal subcarriers*. The following sections will detail these concepts.

3.6.1 Complex-Valued Waveforms

Most modern standards describe the modulated waveform with a complex-valued baseband signal model. The **signal model** is an analytic expression for the waveform. Equation (3.2) provides a generic analytic expression for a complex-valued baseband signal model. The signal model in Eq. (3.2) is **complex-valued**, meaning it has both real and imaginary components. The signal model is also at **baseband**, meaning that it has a center frequency of 0 Hz. The signal model in Eq. (3.2) is composed of a magnitude, $M(t)$, and a phase, $\theta(t)$. The magnitude and phase of the signal model in Eq. (3.2) are time-varying and possibly independent. The magnitude is always a positive value.

$$s(t) = M(t)e^{j\theta(t)} \tag{3.2}$$

The use of a complex-valued representation offers several benefits. One benefit is that the sampled bandwidth is doubled since positive and negative frequencies

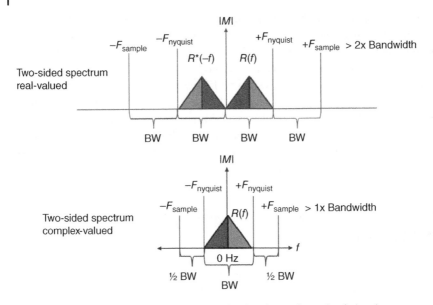

Figure 3.6 Two-sided spectrum of a real-valued and complex-valued signal.

now carry unique information. This concept is illustrated in Figure 3.6. The spectrum of a real-valued signal has a conjugate image in the negative frequency range. This is called *conjugate symmetry* or *Hermitian symmetry*. The spectral components in the negative frequency range do not contain any unique information in the case of a real-valued signal as those components are conjugate-symmetric images of the components in the positive range. The total bandwidth of unique spectral components in Figure 3.6 is labeled *BW*. The fact that the unique spectral components of the real-valued signal are repeated in the negative frequency range causes the effective two-sided bandwidth of the real-valued signal to be twice BW. A complex-valued signal does not exhibit conjugate-symmetry. This means that only one bandwidth of unique spectral components is required to represent the signal. Therefore, a complex-valued signal requires one-half the sampling rate of a real-valued signal for the same unique bandwidth. There is some nuance to the statement that a complex-valued signal requires half the sampling rate of a real-valued signal for the same unique bandwidth. First, this assumes that carrier frequency and phase of the real-valued signal has perfect synchronization between the transmitter and the receiver. Figure 3.7 illustrates the effect of a carrier frequency offset in bringing a real-valued signal down to baseband and a complex-valued signal down to baseband. The real-valued signal has negative frequency components that move up in frequency as the positive frequency components move down in frequency. This causes the real-valued signal to interfere with itself

Figure 3.7 Carrier offset errors in the case of
real-valued and complex-valued signals.

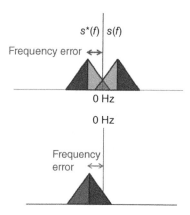

if a frequency offset causes it to move past 0 Hz. To resolve this issue, it is common
for a real-valued system to operate at an *intermediate frequency* (IF), well away
from 0 Hz, to prevent this scenario. The concept of a real-valued signal at an intermediate frequency is illustrated in Figure 3.8. This intermediate frequency adds to
the total amount of bandwidth that is needed to represent the real-valued signal.
The complex-valued signal has no mirror image in the negative frequencies and
can be freely moved around 0 Hz as a synchronization system resolves a carrier
offset.

The sampling rate must be greater than twice the bandwidth of the signal being
sampled. This measure of bandwidth will include the intermediate frequency at
which a real-valued signal is operating. The effect is that the actual sampling rate
necessary for a real-valued signal, given operation at IF, must increase to accommodate that center frequency. Therefore, given that a real-valued signal will operate at an intermediate frequency, the required sampling rate for a real-valued

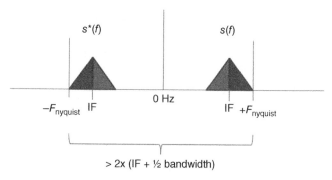

Figure 3.8 Bandwidth of a real-valued signal at an intermediate frequency.

signal is greater than twice the complex-valued baseband signal. Unfortunately, systems are never perfectly synchronized, and the use of an intermediate frequency in a real-valued system is a necessity in many cases. A strong advantage of using complex-valued representation in digital signal processing is that an intermediate frequency is not required. Advantages and disadvantages of direct-conversion front-ends will be discussed in Section 3.7.

Implementing a complex-valued system in the digital domain is trivial, requiring only an implementation of a complex-valued data type. Implementing a complex-valued representation of a signal in the analog domain is nontrivial. Issues encountered when implementing complex-valued signals in the analog domain will be explored in the transceiver section of this chapter.

3.6.2 Symbol Mapping and Linear Modulation

Digital data are a serial stream of bits, where each bit holds a value of one or zero. A wireless communication must map those bits to values to be transmitted over the wireless link. The simplest method would be to take the values as they are, i.e., zero and one. When used to modulate a carrier, the value of zero would nullify the carrier and the value of one would allow the carrier to pass. This type of modulation is therefore called On–Off Keying (OOK). There are a number of inefficiencies with this approach. The next simple scheme is to use the values positive-one and negative-one. The intention is switching the sign of the amplitude of the carrier to reflect the data value being transmitted. This is called binary phase shift keying (BPSK). It is noted that the phase of the carrier is not being directly modulated. BPSK is a **linear modulation** scheme. A linear modulation scheme is one in which a value is directly multiplied with the amplitude of the carrier.

In OOK, the digital data are taken as given, i.e., ones and zeros. In BPSK, the data are mapped to positive-one and negative-one. To use the value of the digital data to generate a new value to be applied to the carrier is called **symbol mapping**. The value to be applied to the carrier is called a **symbol**. In the case of BPSK, the digital data have a value of one or zero, and the symbols have values of positive-one and negative-one.

The individual bits need not be mapped to symbols one at a time. The digital bits can be grouped together and transmitted as one symbol. For example, consider a communication system which transmits two bits in every symbol. The individual bits are grouped in pairs. Interpreted as a binary number, those pairs have decimal values from zero to three. These four data values can then be mapped to four *symbol values* in the complex plane. The process is illustrated in Figure 3.9. A stream of digital data, represented as bits, is grouped. As shown in Figure 3.9, this grouping can be seen as a serial to parallel conversion. The output will be groups, or *frames*, of data. In the case of mapping two bits to a symbol, the grouping is in pairs.

Groups of bits

Figure 3.9 Symbol mapping.

The groups of bits can then form an address into a lookup table (LUT). The LUT stores complex values representing the modulation symbols.

3.6.3 Orthogonal Subcarriers

The concept of orthogonal waveforms is expressed in Eq. (3.3). Two waveforms are said to be orthogonal over some period of time T if the integral of the product of one waveform and the conjugate of the other waveform over time T is zero. The two waveforms in Eq. (3.3) are represented as ψ_0 and ψ_1. Complex conjugation is represented by $(\ )^*$. Because the integral of their product over period T is zero, ψ_0 and ψ_1 are orthogonal.

$$\int_{-\frac{T}{2}}^{\frac{T}{2}} \psi_0(t)\psi_1{}^*(t)\mathrm{d}t = 0 \tag{3.3}$$

Examples of orthogonal waveforms are sine and cosine. A proof is provided as Eq. (3.4) where ω is equal to $\frac{2\pi}{T}$. Conjugation does not matter because both sine and cosine are real-valued. The reason that sine and cosine form an orthogonal set is because they are 90° out of phase with each other. If one thinks of them as vectors, sine is perpendicular to cosine.

$$\int_{-\frac{T}{2}}^{\frac{T}{2}} \sin(\omega t)\cos(\omega t)\mathrm{d}t = \int_{-\frac{T}{2}}^{\frac{T}{2}} \left[\frac{1}{2}\sin(0) + \frac{1}{2}\sin(2\omega t)\right]\mathrm{d}t = 0 \tag{3.4}$$

Sinusoids at different frequencies can also be orthogonal over some period T seconds. If the sinusoids are spaced in frequency at intervals of $\frac{2\pi}{T}$ (in radians/second), the sinusoids will be orthogonal to one another regardless of the phase of the sinusoid. This orthogonality of nonphase coherent sinusoids provides the Fourier basis functions. A proof using complex exponentials is provided in Eq. (3.5). The lack of phase coherence is represented with the phase offset φ (radians), k is any integer, and m is any nonzero integer. The phase offset is constant over the period T. The fact that the phase offset is constant over the period is an important detail that will be used later.

$$\int_{-\frac{T}{2}}^{\frac{T}{2}} e^{jk\omega t} e^{-j(k-m)\omega t + j\varphi} dt = e^{j\varphi} \int_{-\frac{T}{2}}^{\frac{T}{2}} e^{jm\omega t} dt = 0 \tag{3.5}$$

The integration in Eq. (3.5) can be discretized in time to the summation shown in Eq. (3.6). Units of time are now represented by sample index n which spans a range $[0, N-1]$. The integer m is constrained to a value within the range $[1, N-1]$. The integer k is constrained to a value within the range $[0, N-1]$.

$$\sum_{n=0}^{N-1} e^{jk\frac{2\pi n}{N}} e^{-j(k-m)\frac{2\pi n}{N} + j\varphi} = 0 \tag{3.6}$$

The summation in Eq. (3.6) forms the basis for the discrete Fourier transform (DFT). The DFT and the inverse discrete Fourier transform (IDFT) are shown in Eqs. (3.7) and (3.8), respectively. The Fourier basis functions are orthogonal waveforms with spacing $\frac{2\pi}{N}$ (radians).

$$X[k] = \sum_{n=0}^{N-1} x[n] e^{-j\frac{2\pi nk}{N}} \tag{3.7}$$

$$x[n] = \sum_{k=0}^{N-1} X[k] e^{j\frac{2\pi nk}{N}} \tag{3.8}$$

The DFT takes N samples and performs N multiplications per sample. This means the DFT costs N^2 multiplication operations. In a famous paper [14], researchers Cooley and Tukey described a generic algorithm to compute the DFT using only $N \log_2 N$ multiplication operations. This method of computing the DFT with reduced computational cost is called the fast Fourier transform (FFT). The IDFT can also be computed with the same reduction in computational cost and this is referred to as the inverse fast Fourier transform (IFFT).

3.6.4 Modulating the Subcarriers

Equation (3.8) can be seen a collection of orthogonally spaced subcarriers. Each harmonic $e^{j\frac{2\pi nk}{N}}$ is a subcarrier. The variable k ranges from $[0, N-1]$, thus providing N subcarriers. N symbol values will be used to linearly modulate these N subcarriers. Consider the case illustrated in Figure 3.10. For simplicity, the number of subcarriers in this example is 4. Also for simplicity, the symbols used to modulate the subcarriers are either 1 or 0. The imaginary portion of each complex-valued baseband subcarrier is shown. The output signal is a composite of the modulated subcarriers. The symbol of the first subcarrier, located at 0 Hz, is 0 so that subcarrier is nulled eliminating the DC component of the time domain signal. The next

Figure 3.10 Simplified subcarrier modulation diagram.

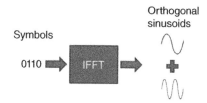

two subcarriers are modulated with a symbol value of 1, so those two subcarriers appear in the composite signal. The composite signal is the sum of all modulated subcarriers.

The benefit of using this IFFT technique as opposed to modulating independent subcarriers is in computational complexity. A system which modulates independent concurrent subcarriers is illustrated in Figure 3.11. The signals are discrete with n being the sample index. Individual symbols within each frame modulate concurrent and independent subcarriers. Multiplying individual subcarriers with N symbols for N samples would require N^2 multiplication operations.

Each subcarrier is modulated with only one complex-valued symbol for every N output samples. The N output samples require an amount of time T seconds equal to the N times the sample period T_s. This means that all subcarriers are spaced $\frac{2\pi k}{T}$ (radians/second) apart from one another where k is an integer. This spacing is unaffected by the phase offset φ of any of the subcarriers as shown in Eq. (3.5) so long as that phase is constant over the period T. This spacing is also unaffected by the magnitude of any subcarrier so long as that magnitude is constant over the period T. If the phase and magnitude of the complex-valued symbols used for

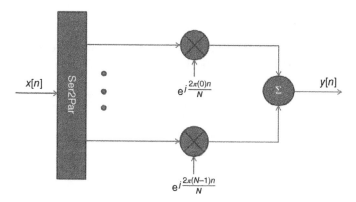

Figure 3.11 Modulating concurrent independent subcarriers.

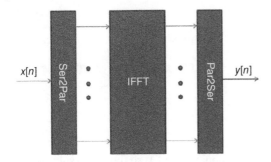

Figure 3.12 Modulating subcarriers using an IFFT.

modulation are constant over the period T ($=NT_s$), then the orthogonality of the subcarriers is maintained.

The same process using an IFFT is illustrated in Figure 3.12. Using an IFFT for this purpose requires only $N \log_2 N$ multiplication operations.

It is by the process illustrated in Figure 3.12 that the orthogonal subcarriers are efficiently modulated with complex-valued symbols. This is **Orthogonal Division Frequency Multiplexing** (OFDM). Multiplexing will be described, and contrasted with *multiple access*, in more detail in Chapter 4.

An expression for OFDM is shown in Eq. (3.9), where $\vec{c}$ is a vector of complex-valued symbols of length N. Modulating orthogonal subcarriers using $\vec{c}$ produces a new vector $\vec{o}$ of length N. This entire vector $\vec{o}$ is referred to as a single **OFDM Symbol**. The OFDM symbol carries N **data symbols**. The creation of complex-valued symbols from data was described in Section 3.6.2.

$$o[n] = \sum_{k=0}^{N-1} c[k] e^{j\frac{2\pi nk}{N}} \tag{3.9}$$

3.6.5 Assigning the Subcarriers

Not all subcarriers need to be modulated with complex-valued symbols. Consider Figure 3.13 from the IEEE 802.11 standard [2]. The standard uses 64 subcarriers in

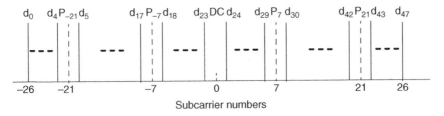

Figure 3.13 IEEE 802.11 subcarriers. *Source:* IEEE 802.11 [2].

Preamble	OFDM data symbol	OFDM data symbol	OFDM data symbol

Figure 3.14 OFDM packet structure.

an OFDM symbol. Only 53 subcarriers are shown in Figure 3.13. The subcarriers not shown are null, meaning that those subcarriers are mixed with zero, and thus are described as **null subcarriers**. The subcarriers not shown are on the edge of the 802.11 channel bandwidth. Leaving these subcarriers *null* allows the OFDM symbol to more easily meet the *spectral mask* for the standard.

The subcarrier in the center of the channel bandwidth is also null. This is done to allow for the use of inexpensive transceivers. This concept will be addressed in Section 3.7.

In addition to the missing subcarriers, which have been nulled, four of the subcarriers are designated as *pilots* in Figure 3.13. These pilot subcarriers are always modulated with a known repeating sequence. The benefit of this is that the receiver can use the pilots for fine synchronization or to update the channel estimate. A packet consisting of some number of OFDM symbols is shown in Figure 3.14. The packet begins with a preamble. This *preamble* will contain known data to help the receiver perform coarse synchronization and calculate a coarse channel estimate for equalization. As time goes on, the pilots in the individual OFDM data symbols allow the receiver to fine-tune those estimates.

3.6.6 Further Reading on OFDM

The goal of this section is to provide the reader with enough information to understand the OFDM modulation scheme, which is used by most of the standards explored in this book. The work of Schmidl and Cox [15] is a seminal reference on the coarse synchronization alluded to in Section 3.6.5.

3.7 Direct-Conversion Transceivers

To understand the design decisions made in the air interfaces defined by the standards discussed in this book, it is necessary to review concepts relevant to modern transceivers. Among these concepts is the advent of *direct-conversion*. While the concept of the direct-conversion is very old, the direct-conversion transceiver became practical in the 1990s [16]. The basic concept of the direct-conversion transceiver, and more importantly, its impact on OFDM symbols, will be discussed here. Extensive details on direct-conversion can be found in References [17] and [18].

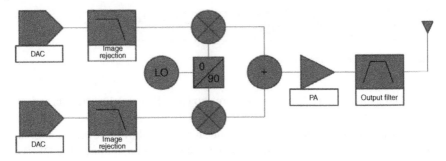

Figure 3.15 Direct-conversion transmitter.

A block diagram of the radio-frequency components within a direct-conversion transceiver is shown in Figures 3.15 and 3.16. Figure 3.15 shows a simplified block diagram for a direct-conversion transmitter and Figure 3.16 shows a simplified block diagram for a direct-conversion receiver. A direct-conversion transceiver is called so because it skips any intermediate frequency stages, bringing a signal directly from or to baseband (0 Hz center frequency) during the analog mixing stages. The **direct-conversion receiver** (DCR) directly downconverts a signal to 0 Hz and a **direct-conversion transmitter** directly upconverts a baseband signal (centered at 0 Hz) to the desired transmit center frequency.

This architecture takes advantage of Euler's formula (Eq. 3.10) to create a complex-valued oscillator and mixing stage.

$$e^{j\theta} = \cos(\theta) + j\sin(\theta) \tag{3.10}$$

This architecture has become popular with the advent of radio-frequency-integrated chips (RFIC). Direct-conversion techniques offer twice the instantaneous bandwidth as compared to heterodyning techniques. However, this comes at a

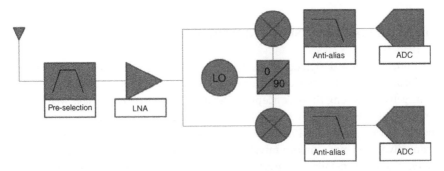

Figure 3.16 Direct-conversion receiver.

cost. Due to imperfect analog component values, there exists an imbalance in the cosine and sine generators in the direct-conversion paths which results in problems, such as local oscillator (LO) leakage and conjugate Images.

One consequence of the impairments in the direct-conversion transceiver is that the center of the transmitted (or received) bandwidth is corrupted by a copy of the local oscillator (*leakage*). A common method of coping with this in an OFDM system is to leave the center subcarrier (0 Hz) as *null* meaning there is no information there. Leaving the center subcarrier blank decreases the total potential throughput of the OFDM symbol by one subcarrier. The trade-off is that this prevents the need for more expensive radio hardware.

References

1 (2019). IEEE Standard for Definitions and Concepts for Dynamic Spectrum Access: Terminology Relating to Emerging Wireless Networks, System Functionality, and Spectrum Management, *IEEE Std 1900.1-2019 (Revision of IEEE Std 1900.1-2008)*, 1–78 (23 April 2019). doi:10.1109/IEEESTD.2019.8694195. https://ieeexplore.ieee.org/document/8694195.

2 *IEEE Standard for Information technology—Telecommunications and information exchange between systems Local and metropolitan area networks—Specific requirements - Part 11: Wireless LAN Medium Access Control (MAC) and Physical Layer (PHY) Specifications*, in IEEE Std 802.11-2016 (Revision of IEEE Std 802.11-2012) , vol., no., pp.1–3534, 14 Dec. 2016.

3 *IEEE Recommended Practice for Information technology – Local and metropolitan area networks – Specific requirements – Part 15.2: Coexistence of Wireless Personal Area Networks with Other Wireless Devices Operating in Unlicensed Frequency Bands*, in IEEE Std 802.15.2-2003 , vol., no., pp.1–150, 28 Aug. 2003.

4 IEEE Standard for Information Technology – Local and Metropolitan Area Networks – Specific Requirements – Part 22: Cognitive Wireless RAN Medium Access Control (MAC) and Physical Layer (PHY) Specifications: Policies and Procedures for Operation in the TV Bands. *IEEE Std 802.22-2011*, 1–680 (1 July 2011). doi:10.1109/IEEESTD.2011.5951707. https://ieeexplore.ieee.org/document/5951707.

5 Cordeiro, C., Challapali, K., and Birru, D. (2006). IEEE 802.22: an introduction to the first wireless standard based on cognitive radios. *IEEE Journal of Communications* 1 (1): 38–47.

6 Sklar, B. (2001). *Digital Communications: Fundamentals and Applications*. Upper Saddle River, NJ: Prentice Hall.

7 Rappaport, T.S. (2002). *Wireless Communications: Principles and Practice*. Upper Saddle River, NJ: Prentice-Hall.

8 Sklar, B. (1997). Rayleigh fading channels in mobile digital communications systems Part I: characterisation. *IEEE Communications Magazine* 35 (9): 136–146.

9 Chayat, N. (1997). *802.11-97/96 Tentative Criteria for Comparison of Modulation.* IEEE 802 LAN/MAN Standards Committee.

10 Erceg, V. et al. (2004). *IEEE 802.11-03/940r4 TGn Channel Models.* IEEE 802 LAN/MAN Standards Committee.

11 Breit, G. et al. (2009). *IEEE 802.11-09/0308r12 TGac Channel Model Addendum.* IEEE 802 LAN/MAN Standards Committee.

12 M.A. Rahman et al. (2010). IEEE 11-10-0154-00-00af Channel Model Considerations for P802.11af. IEEE 802 LAN/MAN Standards Committee.

13 Sofer, E. and Chouinard, G. *Chouinard, WRAN Channel Modeling, IEEE 802.22-05/55r7. IEEE 802 LAN/MAN Standards Committee.* IEEE 802 LAN/MAN Standards Committee.

14 Cooley, J.W. and Tukey, J.W. (1965). An algorithm for the machine calculation of complex Fourier series. *Mathematics of Computation* 19 (90): 297–301.

15 Schmidl, T.M. and Cox, D.C. (1997). Robust frequency and timing synchronization for OFDM. *IEEE Transactions on Communications* 45 (12): 1613–1621.

16 Abidi, A. (1995). Direct-conversion radio transceivers for digital communications. *IEEE Journal of Solid-State Circuits* 30: 1399–1410.

17 Tsui, J. (2001). *Digital Techniques for Wideband Receivers.* Raleigh, NC: SciTech Publishing.

18 Parssinen, A. (2001). *Direct Conversion Receivers in Wide-Band Systems.* Boston, MA: Kluwer Academic Publishers.

4

Mitigating Contention in Equal-Priority Access

This chapter will focus on strategies deployed to mitigate contention between emitters of equal priority. In order to address this concept several terms must be defined. The first is how the spectrum can be divided into discrete resource units. The next is what happens when there is contention for those resource units (interference and collision). Once the consequences of contention have been defined, this chapter will revisit the concept of tiers of users. The concept of multiple tiers of users was first addressed in Chapter 1. The concept of multiple tiers of users was further expanded in Chapter 2 in the discussions on regulatory changes that opened TV White space for use, and the changes that instantiated the Citizens Broadband Radio Service Band (CBRS band). This concept of multiple tiers of users will be briefly revisited here for the purpose of establishing a need for contention-mitigation even within a seemingly exclusive-access bandwidth. This will establish the need for contention mitigation among equal priority users and give specific examples of that contention.

After establishing the need for contention mitigation for equal priority users, this chapter will explore several methods of mitigation with specific references to existing wireless coexistence standards. This chapter will focus on multiple-access and spread-spectrum techniques as methods of contention mitigation. These techniques as used in the wireless coexistence standards are different from standards focusing on exclusive-use. Many parameters in standards focusing on exclusive-use scenarios are statically determined, whereas in standards focusing on wireless coexistence parameters are dynamically determined depending upon the current spectrum environment.

Wireless Coexistence: Standards, Challenges, and Intelligent Solutions, First Edition.
Daniel Chew, Andrew L. Adams, and Jason Uher.
© 2021 The Institute of Electrical and Electronics Engineers, Inc.
Published 2021 by John Wiley & Sons, Inc.

4.1 Designating Spectrum Resources

There is only one spectrum, and everyone must share it. Wireless devices broadcast data using the electro-magnetic spectrum as a medium. In order to accommodate multiple concurrent emitters, the spectrum can be partitioned into a set of accessible resources. Each device is going to need access to such a **spectrum resource** in order to communicate. The exact definition of a spectrum resource is contextual. A definition of a *channel* must be established in order to better understand this. Chapter 3 provided definitions for different usages of the term "channel." Two of those usages are important to this discussion. Those are *frequency channel* and *logical channel*. Those two will be quickly recapped here:

- A **frequency channel** is a specific center frequency and bandwidth.
- A **logical channel** exists in a frequency channel and will be differentiated from other logical channel in that frequency channel by some means such as time slots or code.

The most basic definition of a spectrum resource is a frequency-channel allocated to an entity for use at a specific geographic location and constrained in transmitting power. By way of propagation constraints as discussed in Chapter 3, a given emitter has a limited range. Therefore, a frequency channel allocated to one emitter can be allocated to another emitter concurrently if those two emitters are out of range (sufficiently far apart). This adds a geographic component to the concept of a spectrum resource. A spectrum resource can also be divided in time, where different emitters take turns using the resource. This is sometimes referred to as a *time-frequency resource,* which is defined as a point in time and frequency. A spectrum resource can also mean concurrent use of a given frequency channel by way of code separation as will be discussed in Section 4.12 or time separation as will be discussed in Section 4.10. Common to all these definitions are exclusivity. The user of a spectrum resource expects to have exclusive access to that resource. This exclusivity is meant to limit co-channel interference.

4.2 Interference, Conflict, and Collisions

Noise at the receiver was discussed in Chapter 3. However, noise is not the only thing which muddles reception. Consider trying to have a conversation in a crowded room. It can be difficult to hear the person to whom one is speaking over the din of the crowd. Radios suffer a similar phenomenon often referred to as interference. In order for this multitude of devices to coexist, the devices must keep interference to a tolerable minimum.

Channelization, as first discussed in Chapter 3, is a common means to mitigate contention between multiple users. Channelization can take the form of exclusive use of frequency channels, or exclusive use of time slots on a given frequency, or techniques such as spread-spectrum techniques.

Despite such exclusivity, emitters still interfere with one another. Interference can exist as *adjacent channel interference* (ACI), which occurs with two emitters transmitting in frequency channels that are adjacent. The two emitters will attempt to limit their transmit power to be within the respective bandwidths of their frequency channels. To help minimize this type of interference, wireless standards may specify a *spectral mask* in order to minimize interference in adjacent frequency channels. However, no system is perfect, and two frequency-adjacent emitters will always cause some interference to one another. When two emitters share the frequency channel at the same time, the interference is called *co-channel interference* (CCI).

A **collision** is defined as two nodes transmitting on the same frequency-channel within the same time-period, and with no other mitigating factor. The two transmitted signals inadvertently interfere with one another. This can cause a loss in the wireless link for one or both emitters. A collision is CCI in the sense that the interference takes place in the same frequency and logical channel, however a collision implies that the interference is more destructive and not the result of planned co-use. Collisions will be an important concept when addressing spectrum access in the unlicensed band.

As an alternative to exclusive use, the individual emitters accessing the spectrum can be developed with the ability to intelligently move their transmissions based on observations of the spectrum. This is referred to as Spectrum Sensing and/or Dynamic Spectrum Access. Dynamic Spectrum Access is a technique employed in certain bands by *secondary users*, and will be explored more in Section 4.4. To define what a secondary user is, one must first define what a *primary user* is.

4.3 What Is a Primary User?

Primary Users are defined by IEEE 1900.1 [1] as "Users with higher priority or legacy rights on the usage of a particular spectrum frequency band." This means that a primary user has the right to transmit on a given channel, and any other user must yield.

Primary users are called *incumbent users* in some literature, such as TV White space applications where a broadcast station, or another user designated by the spectrum regulatory authority, has primary rights to a frequency channel. Broadcast stations and their exclusive use of a bandwidth are not the only examples of primary users. Users that are subscribers to a licensed wireless service, e.g. cell-phones, are also primary users. Wireless microphones operating in the TV band also present an interesting case of a primary user. For all unlicensed white space devices, the wireless microphones in the TV band are considered primary users. The secondary users trying to dynamically access white space in the TV band must yield to these wireless microphones. However, the wireless microphones are secondary to the TV broadcast. The wireless microphones have explicit permission from the regulatory authority to operate in those bands provided they do not interfere with broadcast television. The IEEE 1900.1 standard notes this ambiguity in the definition of primary users.

Consider the example of a cellular system. A cellular service provider obtains exclusive use of spectrum resources at a specific geographic location through spectrum regulatory authorities, not unlike the broadcast stations, to provide a service. In the case of cellular service providers, the service is wireless connectivity. One difference between a cellular service and a broadcast station is that in the cellular service there are multiple emitters. A cellular base station provides connectivity to a number of concurrent users. The cellular system shares the spectrum resources to which it has exclusive rights with mobile user equipment subscribed to its service; however, it is not required to share those spectrum resources with any other wireless service. Thus, the cellular base station has complete control to coordinate between subscribed mobile user equipment and determine how those users operate within the licensed spectrum resources. All of those subscribers are also primary users of that frequency channel, even though they are not directly licensed.

Quality of Service (QoS) is an application layer concern. A base station may direct subscribers to use more or less spectrum resources in order to maintain a specified QoS for a given application. For the purposes of defining primary users in the context of this discussion, all the subscribers will be considered equal to one another in their right to access the spectrum resource, regardless of QoS concerns.

4.4 Tiers of Users

In the United States, the Federal Communications Commission (FCC) performed a study and found much of the allocated spectrum was being underutilized [2]. This underutilization opens up the possibility of *spectrum sharing*. The concept allows a class of spectrum users, called **secondary users** (SU) to have access to spectrum otherwise reserved for the *primary users* (PU). Secondary users are of lower priority than the primary users, and in certain bands an SU may be allowed to use a spectrum resource if a PU is not using it.

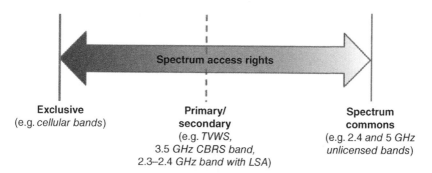

Figure 4.1 The continuum of access models. *Source:* Voicu et al. [3]. (© [2020] IEEE).

These concepts are illustrated in Figure 4.1 from Reference [3]. On the one extreme of Figure 4.1 are "Exclusive Rights" which presumes complete separation in frequency between concurrent wireless systems. On the other side of Figure 4.1 is *Spectrum Commons*, referring to unlicensed use, where various wireless systems are free to transmit anywhere in the designated band and all are expected to mitigate any interference on their own.

The manner in which secondary users coexist with primary users is beyond the scope of this chapter. Chapter 5 will address these mechanisms. What is important in this chapter is to establish the existence of these tiers and discuss how contention between equal peers is mitigated.

4.5 Unlicensed Users

The spectrum regulatory authorities of different nations have allocated certain segments of the spectrum for unlicensed use. One internationally recognized unlicensed band is the 2.4 GHz Industrial, Scientific and Medical (ISM) band. The 2.4 GHz ISM band is available for unlicensed use in most countries. A product designed to operate in the 2.4 GHz band may be sold as-is across international markets. There are some small variations in the definition of the 2.4 GHz ISM band from nation to nation. An example of this variation is that the United States recognizes a slightly smaller 2.4 GHz ISM than other countries and thus the IEEE 802.11 channels 12 and 13 are not available for use within the United States. Some unlicensed bands are specific to individual nations. The Short Range Device band (SRD) is an unlicensed band in Europe and covers a bandwidth in the 800 MHz range. The United States offers an unlicensed ISM band in the 900 MHz range.

4.6 Contention in Spectrum Access and Mitigation Techniques

In the aforementioned example of a cellular system, every node in a given cell will transmit concurrently in order to allow users to have a phone call in which they can speak and hear at the same time. This creates a contention for spectrum access between the sending information to the base station and receiving information from the base station. *Duplexing* is the means of mitigating this specific form of contention to allow concurrent (*or emulated concurrency*) transmission and reception, and this concept will be discussed in Section 4.8. The classical techniques for duplexing will be summarized in Section 4.8 to provide some background and context. More recent trends in duplexing will also be addressed specifically citing examples in wireless coexistence standards which stress adaptive approaches.

In a cellular system, the base station and all mobile subscribed units transmit. The subscribed users come and go, accessing the wireless network on demand. The Medium Access Layer (MAC) must provide some means of contention mitigation to address this contention among equal users. This situation is not exclusive to primary users. Secondary users also have this problem. The secondary users of a spectrum resource are required to yield to a primary user; however, secondary users need not yield to each other. This creates a potential for contention in any frequency channel used if no means of coordination are employed. In the wireless coexistence standards discussed in this book, more flexible approaches to multiple access have been devised that allow not just individual users to share a frequency channels but entire networks to share frequency channels. This chapter will provide a summary of classic multiple access techniques and address emerging adaptive approaches introduced in the wireless coexistence standards.

The issue of which users have the right to access a spectrum resource becomes more muddled in the case of unlicensed bands. None of the users operating in these unlicensed bands hold legal priority over any other. Therefore, wireless devices in these regions must deploy their own interference mitigation.

4.7 Division of Responsibility among the Protocol Layers

The various duties that enable spectrum sharing and contention mitigation are assigned to different layers of the protocol stack. This idea was first addressed in Chapter 1 when discussing the role of standardization in wireless coexistence. This concept will be briefly revisited here to keep the idea fresh in the reader's

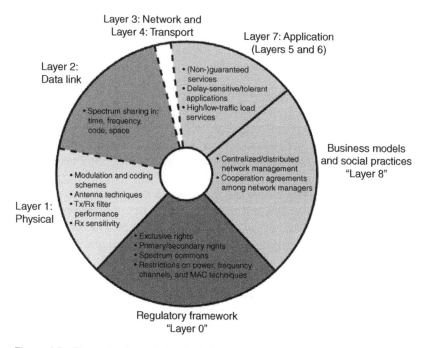

Figure 4.2 The technology circle of wireless coexistence. *Source:* Voicu et al. [3]. (© [2020] IEEE).

mind. Figure 4.2 from Reference [3] illustrates this division of responsibility. Reference [3] provides a survey of spectrum sharing, specifically across different technologies and organizes them into a *technology circle* based on the seven-layer Open Systems Interconnection (OSI) stack. Two additional layers have been added, those being *Layer 0* representing government regulations, and *Layer 8* representing business interests. Layer 0 defines the rules for accessing some band in the spectrum. The regulatory bodies may declare a band to be unlicensed, or may declare a band for exclusive use. This layer 0 is below the layer 1, the physical layer, and drives design decisions of that physical layer. For example, if the layer 0 defines the band as unlicensed leaving each system to fend for themselves, then the physical layer must be designed with interference mitigation in mind. However, should the band be declared for exclusive use, then no such interference mitigation consideration needs to be made. On the other side of layer 0 is layer 8. Layer 8 represents standard practices of the users of wireless systems. Layer 7, which is the application layer, provides input into the business model which is layer 8. The needs of business drive the regulations, thus layer 8 drives layer 0 producing a circular structure.

This chapter is focused on two of these layers:

- Physical (Layer 1), and
- Medium Access (a sublayer of Layer 2).

Layer 0 determines whether or not multiple tiers of users exists, at what range wireless systems may operate, and will restrict design decisions. Regulatory authorities make such determinations and those rules were discussed in Chapter 2. Layer 1, called the *Physical Layer*, may employ spectral masks and spread spectrum techniques to enable wireless coexistence. The use of spectral masks may be initially driven by regulation, as the relationships in Figure 4.2 suggest, but wireless standards will often define spectral masks for their own needs within the bounds of regulation as will be shown in Section 4.11. If a wireless system is deployed in a band where one assumes that interference will happen, then the system may employ a mitigation strategy and seek to minimize the effect of that interference. Such mitigation in Layer 1 makes the wireless link more robust a priori. Layer 2 contains the Medium Access Control (MAC) sublayer and this determines duplexing and multiple access techniques. It is the MAC sublayer that determines how numerous nodes in the wireless network can be assigned to different spectrum resources, either a priori or dynamically, by some central authority.

4.8 Duplexing

One of the first issues of contention to resolve between spectrum access for equal-priority users is how each user can receive and transmit concurrently. Consider the application of a phone call. During a phone call, the users expect to speak and hear at the same time. Contrast that with a pager system. In a pager system, the individual pager node monitors the shared channel waiting for a message and the base station sends messages. These two applications have different needs. One needs each node to at least emulate concurrent receive and transmit. The other works in a single direction.

Duplexing describes the ability for a node in a network to both transmit and receive. The idea is "two-way communication;" that the node can talk and listen at the same time. There are three general categories to duplexing: Simplex, Half-Duplex, and Full-Duplex.

A simplex system is a one-way link [4]. One node is always the transmitter and one node is always the receiver. A Broadcast TV station transmitting to a television is an example of a simplex system.

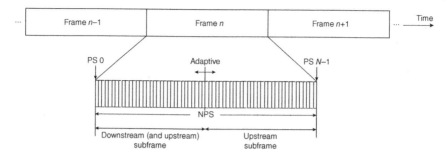

Figure 4.3 802.22 TDD. *Source:* IEEE Std 802.22-2011 [5].

Some wireless systems allow multiple nodes to transmit, but not concurrently. If nodes in a network take control of the shared channel when they transmit, this can be described as a half-duplex system [4]. Wi-Fi is a half-duplex system. Within a given basic service set, only one node may transmit on the shared frequency channel at any time.

If all nodes in a network can receive and transmit data concurrently, then that is called a full-duplex system [4]. In order to establish a full-duplex system, both parties on the link must be able to transmit and receive at the same time.

There are several methods to establish this concurrency. Among the methods of duplexing are Frequency Division Duplexing (FDD) and Time Division Duplexing (TDD).

In a FDD system, nodes in the network transmit and receive on different frequency channels. In a FDD system, the nodes are truly full-duplex in that they transmit one frequency and receive on another, fully concurrently. In a TDD system, nodes transmit and receive on the same frequency channel, but at different times. TDD thus emulates a full-duplex system even though the nodes do not transmit and receive at the same time. IEEE 802.22 is an example of a TDD system, and is illustrated in Figure 4.3 from [5].

4.9 Multiple Access and Multiplexing

Both "multiplexing" and "multiple access" schemes allow multiple users transmit data concurrently within one band. The process of allowing multiple signals across one physical resource is called "multiplexing." Allocating these limited resources to independent users is called "multiple access." Multiplexing implies that the signals are combined together through a coordinator and that the method of combining the data is static. In multiplexing, all multiplexed signals may come from the same user. Multiple Access implies there is more than one individual user accessing the shared medium. Multiple Access also implies a more dynamic allocation of the shared medium.

In Time Division Multiplexing (TDM), multiple signals are allocated time slots and transmitted together. In Frequency Division Multiplexing (FDM), multiple signals are allocated to different frequency channels within a shared band and then transmitted concurrently. An example of Frequency Division Multiplexing can be found in the baseband spectrum of Broadcast FM.

Multiple access in time is called Time Division Multiple Access (TDMA), and in frequency is called Frequency Division Multiple Access (FDMA). TDMA and FDMA divide use of the common medium as do TDM and FDM respectively.

The distinction between multiplexing and multiple access is more pronounced when discussing the concepts of Orthogonal Frequency Division Multiplexing (OFDM) and Orthogonal Frequency Division Multiple Access (OFDMA). OFDM represents a multicarrier modulation technique. Any one OFDM symbol transmitted is composed of many subcarriers each carrying data. A single node in a network transmits all subcarriers simultaneously. OFDMA, on the other hand, is a multiple access technique allowing multiple nodes to transmit the on a dynamically allocated number of the orthogonally spaced subcarriers at dynamically allocated times.

The primary goal of multiplexing or multiple access techniques is to be able to deconflict common access to a shared medium (the spectrum) allowing multiple sources to send data across that shared medium, and to be able to separate that data at the receiver.

4.10 Frequency and Time Division Multiple Access

Frequency Division Multiple Access (FDMA) and Time Division Multiple Access (TDMA) are two of the oldest and simplest multiple access techniques. This section will detail each, give examples of their use, and show how the two techniques can be combined.

FDMA is a multiple access technique that deconflicts concurrent emitters in a given band by separating them into unique frequency channels. In FDMA, individual nodes have exclusive access to some specified frequency-channel. FDMA was the multiple access system used by the first generation of cellphones (1G). In order to prevent any unintentional overlap between the individual users, FDMA employs guard bands between the channels [6]. Such unintentional overlap is referred to as "adjacent channel interference". Figure 4.4 illustrates an arbitrary FDMA system where different users in a shared band are separated into frequency channels. In this example, there are three users at three individual frequency channels. Guard bands exist between the users to prevent adjacent channel interference.

That Broadcast TV stations have rights to specific frequency channels is an example of FDMA. Trunking in mobile radios is another example of FDMA, and a far more dynamic one than broadcast stations.

A "channel plan" is an example of an FDMA system. For example, Wi-Fi has a channel plan and can separate Access Points (APs) by frequency channel. Consider a scenario in which multiple Wi-Fi Access Points (APs), all within range of each other, and transmitting on different nonoverlapping frequency channels. In this Wi-Fi scenario, each AP forms a Basic Service Set (BSS) with multiple clients. Within any one BSS, the nodes use Carrier Sense Multiple Access to share the frequency channel. The function of FDMA is to segregate the APs and the devices associated with each. Separating those APs onto different nonoverlapping frequency channels prevents contention and interference. Networks using IEEE 802.22 do not employ FDMA, but rather dynamic spectrum access for secondary users.

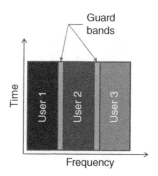

Figure 4.4 Frequency division multiple access.

Time Division Multiple Access (TDMA) is multiple access technique that allows individual users to concurrently transmit into a shared band at different times but on the same frequency channel. Because the frequency channel is shared, the individual users will be divided into "logical channels."

Thus TDMA can present something of a turn-based system. The periods of time over which an individual users can transmit into the shared band is called a time slot. The term "burst" is also common to describe the transmission from any one user. A TDMA system could then be said to be a "bursty" signal as opposed to a continuously transmitted signal. Second generation cellphone systems (2G) like GSM are based on TDMA. The basic transmission in GSM is called a "burst".

Figure 4.5 illustrates an arbitrary TDMA system where different users are grouped into a TDMA frame. The users are independently transmitting and there may be inaccuracies in timing. These inaccuracies could cause overlap between individual bursts. TDMA frames have

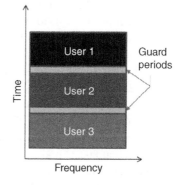

Figure 4.5 Time division multiple access.

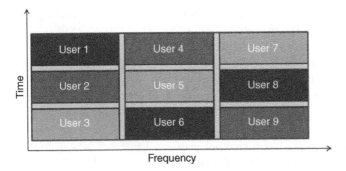

Figure 4.6 FDMA and TDMA.

guard periods to prevent inadvertent transmission collisions due to inaccuracies in timing.

TDMA systems offer several advantages over FDMA systems. First, the TDMA systems can transmit to multiple nodes on one carrier. Having only one carrier amplified at the transmitter mitigates intermodulation distortion. Another benefit is that the TDMA system can allocate new time slots more easily than new frequency-channels and bandwidths can be allocated for a FDMA system (Figure 4.6).

TDMA and FDMA techniques can be used in the same system. The Global System for Mobile Communications (GSM) provides an example of FDMA [7] and TDMA [8] being used together. GSM is a 2G protocol for cellphone service. In the context of this discussion, the mobile devices are all primary users. The primary users, subscribed to a specific service-provider, enjoy exclusive access to their frequency channels. In fact, each GSM base station exists on a pair of specific frequency channels (full duplex) called the *forward channel* (transmitting to the individual users) and the *reverse channel* (transmitting to the base station). The base station does not share its forward channel with any other emitter. Individual users share access to the reverse channel by TDMA. It is important to note that the access method described for these cellphones only guarantees coexistence with nodes in their own specific subscriber network.

The IEEE 802.22 standard does not describe multiple access for primary users; however, the standard contains a TDMA mode for secondary users which is not unlike that of primary users. There are two differences between the TDMA mode described in the 802.22 standard and the version described for GSM. The first difference is that 802.22 nodes must first find white space in the spectrum (as a secondary user) and then coordinate access to that white space in with TDMA. The second difference between IEEE 802.22 and the TDMA system described for cellphones is that in IEEE 802.22 the base stations coexist with other base stations by

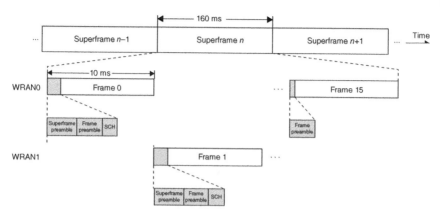

Figure 4.7 IEEE 802.22 TDMA. *Source:* IEEE Std 802.22-2011 [5].

way of TDMA. In the IEEE 802.22 standard this TDMA method between base stations is called "self-coexistence mode" and it only pertains to sharing a frequency channel with other 802.22 systems. The 802.22 systems access a shared white space frequency channel, taking turns on a frame-by frame-basis. This demonstrates a **collaborative coexistence** mechanism. The base stations collaborate and share the spectrum resources.

The 802.22 TDMA process is illustrated in Figure 4.7 from [5]. Each 802.22 base station is allocated some number of frames in a superframe. The 802.22 base stations enjoy exclusive access to the white space frequency channel for that duration of time. The different base stations in Figure 4.7 are represented as Wireless Region Area Network (WRAN) 0 and 1. Time is represented as the horizontal axis, and the two WRANs share access to one white space frequency channel.

The 802.22 coexistence mode does not allow for sharing the white space frequency channel with other non-802.22 systems. Although 802.22 is for secondary users, its TDMA mode assumes exclusive rights. In order to provide collaborative coexistence mechanisms with other non-802.22 networks, some external method is needed. One such external method is the standard IEEE 802.19.1, which will be discussed in more detail in a later chapter.

4.11 Spectral Masks Defined in Standards

A **spectral mask** is a requirement limiting the spectrum of a transmitted signal. Spectrum masks are defined by standards and the primary purpose of a spectrum mask is to limit bandwidth so as to prevent adjacent channel interference. When a wireless signal transmits the actual bandwidth, as an analytic expression of the signal model, may be infinite. Additionally, there are impairments and nonlinearities at the transmitter which cause **spurious emissions** outside the bandwidth of a designated frequency-channel. The term spurious emission implies that the emission is a *spur*, resulting from some harmonic or some analog impairment in the transmitter chain. Any emission outside of the designated frequency-channel bandwidth is considered in error, regardless of the source. The spectral mask imposes a limit on those erroneous emissions.

Chapter 2 discussed spectral masks from a regulatory perspective. Spectral masks are not exclusively defined by regulation and are often defined in wireless communication standards. Regulations are often drivers for physical layer requirements such as spectral masks, but the limitation on bandwidth imposed by the standard may also be set to maximize the number of concurrent primary users connected to a wireless service. The spectral masks for TV white space are an example of this. The width of that mask is established by regional regulation, as will be seen in this section. From there, the standard adds definition for the purpose of addressing some user need such as utilizing two frequency channels at once in order to increase transmission bandwidth and thus increase the total possible data rate.

There are many synonyms to the term "spectral mask". The terminology employed by different wireless standards is inconsistent. The IEEE 802.11 standard refers to its spectral mask as the "transmit mask". The IEEE 802.22 standard refers to its spectral mask as the "RF mask". The terms, "transmit mask" and "RF mask" are to be considered synonymous with "spectral mask."

Spectral masks support wireless coexistence by restricting out-of-band emissions. The spectral mask is one of the oldest techniques for wireless coexistence, as they are designed to keep each wireless signal limited to the bandwidth of a designated frequency-channel.

This section will detail a few examples of spectral masks employed in the standards covered in this book. The intention is to familiarize the reader with practical examples of spectral masks for very different operating environments.

The spectral mask for the Direct Sequence Spread Spectrum (DSSS) PHY in the IEEE 802.11 standard is shown in Figure 4.8 and is intended for use in a 20 MHz frequency-channel in the Industrial, Scientific, and Medical (ISM) band. The mask itself is drawn as flat lines, and underneath is the shape of the magnitude of a *sinc* function. This second *sinc* line is for information only. The magnitude of the spectrum of the DSSS waveform will follow this *sinc* shape. The units "dBr" are "Decibels relative to the reference". The x-axis indicates how many MHz away from the center frequency the measurement is being taken. Thus, 11 MHz from the center frequency, it is expected that transmitted power will be 30 dB down from the reference which is the peak at the center of the transmitted bandwidth.

The spectral mask for the IEEE 802.11 standard OFDM symbols is shown in Figure 4.9. Like the mask in Figure 4.8, the spectral mask in Figure 4.9 is designed

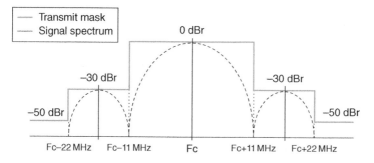

Figure 4.8 Spectral Mask for IEEE 802.11, DSSS. *Source:* IEEE 802.11-2016 [9].

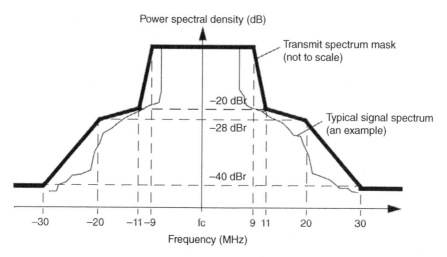

Figure 4.9 Spectral Mask for IEEE 802.11, OFDM 20 MHz. *Source:* IEEE 802.11-2016 [9].

for operation in a 20 MHz frequency-channel in the ISM band. Note that the roll-off is more gradual in Figure 4.9 than in Figure 4.8. This indicates that the *reference level* of the OFDM transmission is lower than that of the DSSS variant. The OFDM transmission is much more spectrally efficient, spreading out the power of the transmission evenly across the multiple subcarriers. The DSSS variant, by contrast, as a large concentration of energy at the center of the bandwidth (which serves as the reference).

An example of a spectral mask for the IEEE 802.11af standard, which is Wi-Fi in TV White space, is shown in Figure 4.10. In this example, two adjacent 6 MHz TV frequency-channels are used concurrently. Each has a spectral mask of its own. The two adjacent channels combine for a concurrent 12 MHz of bandwidth for

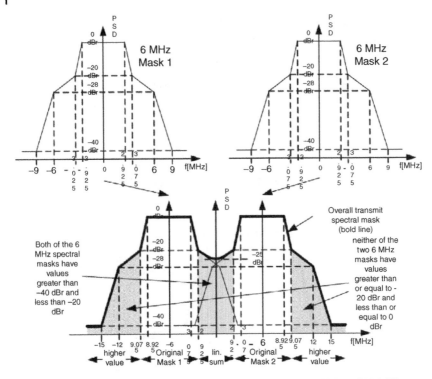

Figure 4.10 Spectral Mask for IEEE 802.11af 6 + 6. *Source:* IEEE 802.11-2016 [9].

the Wi-Fi signal in TV white space. PSD stands for Power Spectral Density. As can be seen, any one 6 MHz channel is expected to use only 5.85 MHz. After that point, a roll-off is expected down to −20 dB relative to the flat peak of the signal PSD. The roll-off continues until ±9 MHz from the center frequency of the channel where any emission is expected to by −40 dB down from the flat peak of the signal PSD. So long as the Wi-Fi signal in TV white only uses one TV white space frequency-channel, that spectral mask is sufficient. When two TV white space frequency-channels are used concurrently, then two spectral masks must be combined as shown in Figure 4.10.

The spectral mask of the IEEE 802.22 standard for operation in a TV white space in the United States is shown in Figure 4.11. The US TV band is divided into fre-quency-channels 6 MHz wide, there for the IEEE 802.22 spectral mask for opera-tion in the US is designed for that bandwidth. The figure has been redrawn here to more clearly enumerate the x-axis. The units "dBc" are "Decibels relative to the carrier". The signal is measured in power spectral density with a bandwidth of 100 kHz. That means that the power measurement is averaged over a 100 kHz

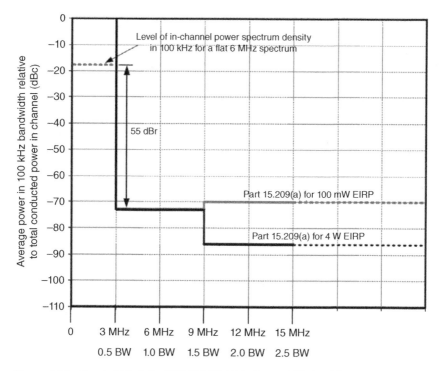

Figure 4.11 Spectral Mask for IEEE 802.22, USA. *Source:* IEEE Std 802.22-2011 [5].

measurement bandwidth. For the first 3 MHz away from the center frequency, the signal is allowed to keep a flat power spectral density, where any measurement averaged over 100 kHz. After 3 MHz, the requirement drop sharply to be 55 dB below the reference of that flat peak in the passband. It is unlikely that an actual signal would drop so suddenly, therefore to meet the requirement, a signal would need to start "rolling-off" before this 3 MHz requirement in order to have 55 dB of attenuation at 3 MHz away from the center frequency.

4.12 Spread Spectrum Techniques

Spread Spectrum techniques provide interference mitigation (*Mitigation Strategy*) as well as resilience to multipath channels. The basic idea behind Spread Spectrum is to apply a reversible process to the desired signal which ultimate spreads the bandwidth of interference, reducing the ill effects of the interference. When operating in unlicensed bands, wireless systems cannot rely on regulation to prevent interference. Therefore these systems must employ their own means to mitigate interference. For example, Bluetooth and WiFi operate in the same 2.4 GHz ISM band. Each is known to interfere with the other. There have been numerous papers written analyzing interference between these two wireless networks in the unlicensed bands, an example can be found in Reference [10]. The two systems have no means to coordinate with one another, and no regulatory authority to prevent co-use of the unlicensed band from causing interference.

Spread spectrum provides a means for a communication system to mitigate the effects of an interferer without the need of coordinating with any other wireless systems. Spread spectrum systems can work without sensing the spectrum to avoid interference. Spread Spectrum therefore provides a **noncollaborative** method of wireless coexistence.

This section will detail two spread-spectrum techniques from the perspective of wireless coexistence. Those are **frequency-hopping** and **direct-sequence** spread spectrum. Spread spectrum will be shown to not only mitigate interference but also provide a means of multiple access, referred to as "Spread Spectrum Multiple Access" in reference [11]. Frequency hopping sufficiently explained to address the concept of **Adaptive Frequency Hopping** (AFH). Direct-sequence spread spectrum will likewise be explained to introduce the **near-far problem** and **successive-interference cancelation**.

4.12.1 Frequency Hopping

The basic idea of frequency-hopping spread spectrum (FHSS) is to change the center frequency of the signal according to some pattern known by the transmitter and the receiver. The benefit of frequency hopping is that a frequency-selective channel or an interferer can adversely affect only part of the transmission.

The transmitter and receiver will "hop" between a set of center frequencies called the **hop set**. The hop set divides the bandwidth over which the FHSS system operates into frequency channels, and defines a time to tune to different frequency channels in that channel plan. Thus, the hop set can be described as a set of time-frequency pairs. *Time* in this sense is periodic. The hop set will be repeated.

Two FHSS systems may operate over the same bandwidth so long as they have different hop sets. Ideally, the two hop sets will not contain any conflicting time-frequency pairs, that is, the two systems will not attempt to access the time-frequency pair. The two systems will be separable from each other in that a receiver for one system will access only one set of time-frequency pairs and ignore all others.

These hop-sets may inadvertently share time-frequency pairs. For example, Reference [12] explores cases where Bluetooth piconets interfere with other Bluetooth piconets. These inadvertent collisions are due to the fact that there are a finite number of time-frequency pairs, and the FHSS systems do not coordinate to prevent using the same pairs.

4.12.2 Adaptive Frequency Hopping

Adaptive Frequency Hopping (AFH) is a variant of FHSS where the hop set can be changed dynamically in response to the environment. The standard IEEE 1900.1 [1] provides a taxonomy of terms relating to wireless coexistence and advanced radio functions. Wherever possible, the terminology defined in that standard will be used in this book. There are cases where some divergence is required. For example, IEEE 1900.1 does not disambiguate between frequency hopping and adaptive frequency hopping. In the context of this book and the concepts detailed herein, the adaptation component of AFH is of enough importance to warrant distinction. The AFH system either *monitors* or *senses* the environment and makes corrections to the hop set to avoid problematic hops.

A classic example of this is the interference between IEEE 802.15.1 Wireless Personal Area Networks (WPAN) [13] and IEEE 802.11 Wireless Local Area Networks (WLAN) [9] in the 2.4 GHz ISM band. IEEE 802.15.1 was the original standard for the lower layers of Bluetooth and IEEE 802.11 is the standard for the lower layers of Wi-Fi. Both Bluetooth and Wi-Fi have become ubiquitous and are often used simultaneously. The IEEE 802.15.1 WPAN frequency hops across 79 channels spanning much of the 2.4 GHz ISM band. One can imagine using a mobile phone connected to a Wi-Fi while also using an ancillary device, such as ear buds, connected to the mobile phone via Bluetooth. The interference between the WPANs and the WLANs became of sufficient concern to warrant the creation of a task group to tackle it. The result of that task group was the standard IEEE 802.15.2 [14].

Among the recommendations in IEEE 802.15.2 was Adaptive Frequency Hopping, to be implemented in IEEE 802.15.1, as a noncollaborative means to mitigate interference. While the IEEE 802.15.2 standard has been officially withdrawn, AFH was adopted into the IEEE 802.15.1 standard. The impact of IEEE 802.15.2 is still seen in the Bluetooth core specifications which recommends AFH as a noncollaborative interference-mitigation technique. The Bluetooth Core specifications do not state how to implement AFH.

A given frequency-hopping WPAN system may employ Bit Error Rate (BER) as a metric to determine the usability of any given frequency channel in the hop-set, where the Bluetooth device will monitor BER as a function of the current frequency channel. If a low BER correlates to the use of a particular frequency channel, that frequency channel will be removed from the hop-set.

This capability is distinct from frequency hopping in that a monitoring function is essential to the operation. The wireless system adapts to the environment with the intention of avoiding interference with other wireless systems in the same band.

4.12.3 Direct Sequence Spread Spectrum and Code Division Multiple Access

Direct-Sequence Spread Spectrum is a technique where the signal is mixed with a pseudo-random sequence of *chips*. This sequence of chips is called the **spreading code**. The chips have a much higher rate than the symbols in the signal. The effect is to spread the energy of the signal across a broader bandwidth. The rate of the symbols id the *symbol rate* and the rate of the chips is the *chip rate*.

The goal of the spreading code is to be able to correlate with itself (*autocorrelation*) and produce low autocorrelation sidelobes. Ideally, the autocorrelation $R_{xx}(k)$ of a spreading code c_x would result in a discrete impulse $\delta(k)$. $c_x{}^*$ represents the complex-conjugate of spreading code c_x. L represents the length of the spreading code. In practice the result is not a discrete impulse but something impulse-like, represented in Eq. (4.1) with $\hat{\delta}(k)$. An example of a length-7 barker code autocorrelated with itself is shown in Figure 4.12.

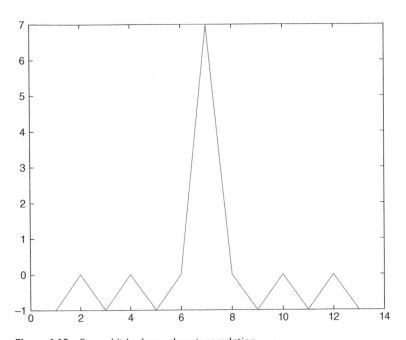

Figure 4.12 Seven-bit barker code autocorrelation.

$$R_{xx}(k) = \sum_{l=0}^{L-1} c_x(l)c_x{}^*(k-l) = \hat{\delta}(k) \qquad (4.1)$$

Another desirable property of these spreading codes is that they be orthogonal with other spreading codes. This would mean that the cross-correlation $R_{xy}(k)$ of two codes c_y and c_x would be zero. In practice the codes are not orthogonal and the cross-correlation process produces **cross-correlation noise**. The cross-correlation process is shown in Eq. (4.2). The cross-correlation noise is represented as $R_{xy}(k)$.

$$R_{xy}(k) = \sum_{l=0}^{L-1} c_y(l)c_x{}^*(k-l) \qquad (4.2)$$

The autocorrelation of Eq. (4.1) and cross-correlation of Eq. (4.2) directly apply to the de-spreading process. In order to de-spread the signal, the complex-conjugate of the spreading code is mixed with spread signal. This product is then accumulated over the symbol period. The length of the spreading code will span the symbol period.

This has several beneficial effects. DSSS can mitigate narrowband interference. When the signal is de-spread, the narrowband interfering signal is spread. This causes the power of the interferer to be spread across a much wider bandwidth determined by the chip rate. Another benefit is mitigating multipath effects. When the signal is despread and accumulated, the accumulation result can reveal an estimate of the power delay profile of the channel. This then leads to a concept called a RAKE receiver. The RAKE receiver was first proposed by Price and Green in Reference [15]. The details are beyond the scope of this discussion.

The cross-correlation shown in Eq. (4.2) comes into play when using these spreading codes as a means for multiple access. Code Division Multiple Access (CDMA) is a DSSS technique used to disambiguate multiple users concurrently occupying the same time-frequency resource. This is to say that users are differentiated by their unique spreading codes. CDMA is common to third Generation (3G) cellphone technologies.

Consider the scenario illustrated in Figure 4.13. There are three users transmitting at the same time on the same frequency channel, but their transmissions are

Figure 4.13 Code division multiple access scenario: locations.

User 1, Code 1

User 2, Code 2

User 3, Code 3

Transmitters

Receiver

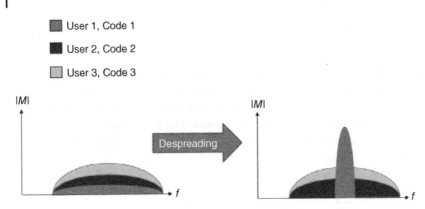

Figure 4.14 Code division multiple access scenario: spectrum.

spread by different codes. The use of these different codes allows the receiver to disambiguate between them. The three users all transmit at the same power; however, the received power for each is different because each user is located at a different distance away from the receiver. This difference in distance will become very important shortly.

The spectrum of the CDMA system from Figure 4.13 is illustrated in Figure 4.14. The different received power levels of each of the users can be seen. The received signal contains all three user signals and is therefore a composite of all three. The signal for user 3 contributes the most to the received power. The signal component for user 3 is the highest power signal component due to the close proximity between user 3 and the receiver. The receiver despreads and demodulates the signal from user 1. When the signal from user 1 is despread, the power received from user 1 is concentrated in a smaller bandwidth yielding a greater power spectral density. Thus, it appears that the signal from user 1 rises above the other users as a function of despreading.

Now consider the spectrum shown in Figure 4.15. Figure 4.15 represents a slight modification in which user 3 is brought even closer to the receiver and thus detected as having even more received power. The received power of user 3 is so great that the receiver can no longer successfully detect user 1. The received power difference between user 1 and user 3 is too large. This is called the **Near-Far Problem**. The codes used to spread and despread are not orthogonal, and there is cross-correlation noise. The power of the cross-correlation noise resulting from applying the despreading code for user 1 to the signal from user 3 is proportional to the received power from user 3. User 3 is too *near* and user 3 is too *far*. The signal from user 3 is too powerful for the receiver to detect user 1.

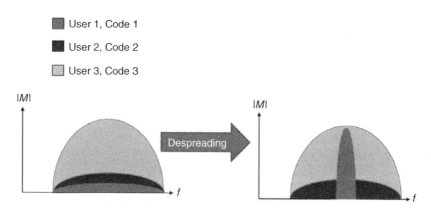

Figure 4.15 Code division multiple access scenario: near far problem.

There are several means to resolve the near-far problem. One is to enable **dynamic power control**. With dynamic power control, a base station can tell a mobile device in a cell to turn down or turn up that mobile device's transmit power. If the base station detects one user as having too much received power relative to the users, as shown in Figure 4.15, the base station can order that too near user to turn down their transmit power and thus convert the scenario in Figure 4.15 to that in Figure 4.14.

Another option is called **success-interference cancelation**. The received signal is a composite of all concurrent user signals. Successive-interference cancelation removes the user components successively and demodulates the user components as they are discovered. The process of successive-interference cancelation demodulates the most powerful user first, then creates an estimated version of that user-signal. The estimate of the most powerful user-signal is slipped 180° out of phase with the received version of that user-signal. The estimated version is then added to the composite received signal. The result is that the estimated user-signal destructively interferes with the most power user-signal removing it from the total received signal. This process is repeated for all users. This technique will be revisited in Chapter 10 when Nonorthogonal Multiple Access is discussed.

4.13 Carrier Sense Multiple Access

Carrier Sense Multiple Access is a multiple access scheme in which users share the entire bandwidth of a given frequency channel. The users are required to first *sense* the channel before using it. Energy detection is typically employed to determine if

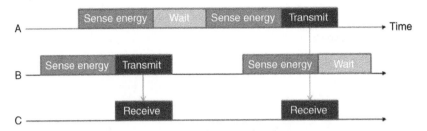

Figure 4.16 Assessing the channel with energy detection alone. *Source:* Chew [16].

the shared channel is in use. Energy detection may be augmented by other detection algorithms. The goal is to determine whether or not the channel is *clear*, meaning there is no other node transmitting on it. If energy in the channel is detected, then the nodes will wait for a pre-determined amount of time before trying again. This wait period will be a random delay. The nodes will sense the channel for a period of time before determining that the channel is clear.

A typical scenario is illustrated in Figure 4.16 from reference [16]. In this scenario, there are three nodes in a network labeled A, B, and C. Figure 4.16 shows the operation of all three nodes in time. Both node A and node B have data to transmit to node C. Both node A and node B will use Energy Detection to determine if the channel is clear. In the scenario illustrated in Figure 4.16, Node B senses for energy in the channel. Sensing that the channel is clear, node B begins to transmit. In this scenario, node A begins sensing the channel shortly after node B. Even though node B began transmission during the sensing period, node A can collect enough energy during the sensing period to determine that the channel is in use. In response to the channel being in use, node A waits for a random back-off period, and then tries again. On the second attempt, node A determines the channel is clear and then begins to transmit. Node B has more data to send to node C, therefore node B sense the channel before transmitting. Node B detects the transmission on the channel, and waits for a back-off period before trying again.

Both the IEEE 802.11 and IEEE 802.22 standards employ channel sensing before transmitting. IEEE 802.11 uses two types of detectors, an IEEE 802.11 preamble detector and an Energy Detector. The IEEE 802.11 preamble detector can provide an improved lower probability of missed detect and a probability of false alarm than the Energy Detector. However, the preamble detector can only detect other IEEE 802.11 transmissions. To sense for any other system using the frequency-channel, the energy detector must be used. IEEE 802.22 follows a similar approach but has many different signal detectors at its disposal. IEEE 802.11af, which is the IEEE 802.11 amendment dealing with operation in TV white space and thus the most comparable to IEEE 802.22, does not include the long list of detectors found in IEEE 802.22. IEEE 802.11af relies on the same two detectors, preamble and energy, defined in previous iterations of the IEEE 802.11 standard.

An Energy Detector is considered a *blind detection method* as it requires no knowledge of the signal it is trying to detect. Using a known preamble for detection is a *data-aided method*.

4.13.1 Collision Avoidance

The problem with the scenario in Figure 4.16 is that the nodes may not successfully detect one another. Consider the case in Figure 4.17 from [16]. The situation is the same except that the timing has changed. With a slight change to the timing, the sensing period does not sufficiently overlap with a transmit period. With such a short overlap in time, not enough energy can be collected and the channel is mistakenly determined to be clear. Both nodes transmit and the result is that neither transmission is successful.

In addition to the problem with timing, there are problems with range that cause CSMA to fail. Consider the diagram in Figure 4.18 from [16]. There are four nodes in this network, A, B, C, and D. The ranges for nodes A and B are illustrated with dashed-line circles around that node. Either node cannot detect nodes outside of their respective circle.

Node A can only detect node B. This means that node A cannot sense the channel for activity from node C, even though both nodes A and C communicate with node B. In that scenario, nodes A and C cannot help but interfere with one another. That scenario is a common problem and is called the *Hidden Node Problem*.

Node A cannot detect node C, and node B cannot detect node D. Therefore is node B transmits to node A and node C transmits to node D, there will be no interference. Unfortunately, node C does not have this information. Therefore, when node C detects node B transmitting, node C assumes it cannot transmit to node D.

Figure 4.17 Collision at the receiver. *Source:* Chew [16].

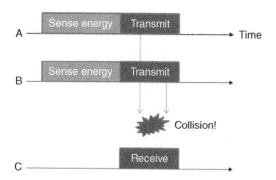

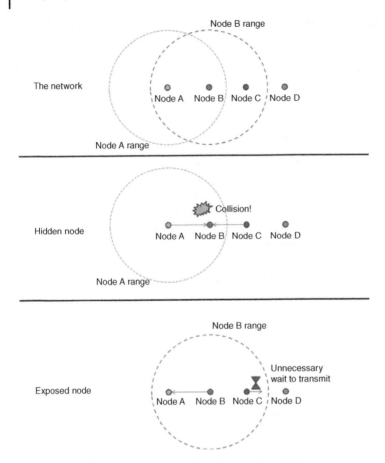

Figure 4.18 Hidden and exposed nodes. *Source:* Chew [16].

The scenario reveals undue contention. Traffic from Node B to Node A has no bearing on traffic from Node C to Node D. This type of problem is called the *Exposed Node Problem*.

In order to resolve these problem, CSMA is augmented to CSMA with Collision Avoidance (CSMA/CA) [17]. A handshaking scheme is implemented to request permission to transmit is created. That handshaking scheme is illustrated in Figure 4.19. The node wanting to transmit sends a *Request-To-Transmit* (RTS) to the desired receiving node. If the channel is clear, the receiving node replies with a *Clear-To-Transmit* (CTS). In the case of the hidden node, Node A can detect the CTS sent from Node B to Node C, and therefore can determine that the channel is occupied. This provides *Virtual Channel Sensing*. For the exposed node problem, if node B detects the RTS but never detects the CTS, it can be determined that the receiving node is out of range. It is important to note that Collision Avoidance requires at least some centralized coordination. Thus, CSMA/CA is not a scheme that allows systems following different standards to share a frequency-channel resource.

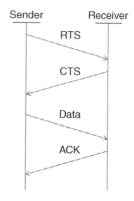

Figure 4.19 Collision avoidance messages. *Source:* Chew [16].

4.14 Orthogonal Frequency Division Multiple Access

Orthogonal Frequency Division Multiple Access (OFDMA) is a multiple access scheme that divides spectrum resources in frequency and time. The OFDMA scheme relies on the multiple subcarriers inside an Orthogonal Frequency Division Multiplexed (OFDM) symbol. An OFDM symbol is a composite of multiple subcarriers, each linearly modulated with a data symbol. In OFDMA, this multicarrier concept is advanced to apply to multiple users. Different users are allocated time slots on the subcarriers within an OFDM symbol. The assignments change in time. The concept is illustrated in Figure 4.20 which shows the assignment of subcarriers to different users. The y-axis is frequency, showing individual subcarriers,

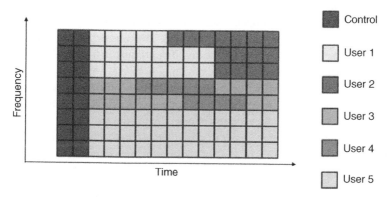

Figure 4.20 Orthogonal frequency division multiple access.

and the x-axis is time, showing individual OFDM symbols. For simplicity, the example shown in Figure 4.20 has eight subcarriers. A column of eight subcarriers composes one OFDM symbol. The OFDM symbols are transmitted in time. The initial OFDM symbols are reserved for control functions. Among the control information will be subcarrier mapping information allowing the end-users to know which subcarriers at what time carry information for them. The remaining OFDM symbols carry information for individual users.

OFDMA offers several advantages over other multiple access techniques. Among these are frequency diversity and adaptive frequency selection.

Adaptive Frequency Selection: An OFDMA system may adapt and select subcarriers based on the frequency selectivity of a wireless channel. Consider that a multipath channel is specific to a pair of nodes and is reciprocal. In some OFDMA systems, such as a Time-Division Duplexed (TDD) system, the nodes share a frequency channel. Because each node is spatially distinct, the multipath channel is unique to the link between any pair of nodes. The reciprocity of the multipath channel means that each of the two nodes will observe the same scattering. OFDMA allows the system to place users in frequencies that minimize fading with respect to the user's unique multipath channel.

Frequency Diversity: An OFDMA system may alternate the subcarriers allocated to multiple users. The effect of this is akin to frequency hopping. The users move to different subcarriers and only part of their transmission is affected by an interferer or by a frequency selective channel.

4.15 Final Thoughts

This chapter focused on methods by which wireless standards provide for coexistence between primary users. Mitigating contention for spectrum access among these users are major design choices in these wireless standards.

Section 4.4 introduced the concept of tiers of users. This was necessary as without tiers of users, then there is no need for the qualifier *primary* in front of the word user. However, this chapter stops short of explaining how users of lower priority coexist with users of higher priority. A common method by which a secondary user may access the spectrum in the absence of a primary user is called *Dynamic Spectrum Access* (DSA). DSA generally is often achieved by having the secondary user *sense* a frequency channel to determine if that frequency channel is occupied. The IEEE 1900.1 standard defines this process of sensing the environment for opportunities as *Opportunistic Spectrum Access* (OSA). In the taxonomy defined in IEEE 1900.1, OSA is an implementation of the larger DSA concept. OSA and the prerequisite sensing algorithms will be discussed in detail in Chapter 5.

The spectrum sensing discussed in Chapter 5 is not limited to secondary users. Carrier Sense Multiple Access schemes, as the name implies, also employ spectrum sensing. CSMA systems often employ energy detection as a *blind search* mechanism, and this will be discussed in Chapter 5. This mechanism is employed because CSMA system operating in unlicensed bands may encounter contention with third-party system using the same frequency-channel. The third-party system is one with which the CSMA standard does not share a wireless standard, and therefore the CSMA system must employ a broader search to avoid collisions with such third-party systems.

The hidden node problem is also something that OSA and CSMA systems share. If the hidden node is a third-party system with which the there is no shared standard, then handshaking cannot resolve that scenario. As will be seen in Chapter 8, collaborative sensing is a means by which an OSA system can overcome the hidden node problem.

References

1 IEEE Std 1900.1-2019 (2019). *IEEE Standard for Definitions and Concepts for Dynamic Spectrum Access: Terminology Relating to Emerging Wireless Networks, System Functionality, and Spectrum Managemen.* IEEE.

2 Federal Communication Commission (2002). *Spectrum Policy Task Force. Report ET Docket 02-135.*

3 Voicu, A.M., Simić, L., and Petrova, M. (2018). Survey of spectrum sharing for inter-technology coexistence. *IEEE Communication Surveys and Tutorials* 21 (2): 1112–1144.

4 Tanenbaum, A.S. (2003). *Computer Networks.* Upper Saddle River, NJ: Prentice Hall.

5 IEEE Std 802.22-2011 (2011). IEEE Standard for Information technology-- Local and metropolitan area networks-- Specific requirements-Part 22: Cognitive Wireless RAN Medium Access Control (MAC) and Physical Layer (PHY) Specifications: Policies and Procedures for Operation in the TV Bands, pp. 1–680, doi: https://doi.org/10.1109/IEEESTD.2011.5951707.

6 Sklar, B. (2001). *Digital Communications: Fundamentals and Applications.* Upper Saddle River, NJ: Prentice Hall.

7 TS 45.005 V16.1.0, 3rd Generation Partnership Project; Technical Specification Group Radio Access Network; GSM/EDGE Radio transmission and reception (Release 16).

8 TS 45.002 V16.1.0, 3rd Generation Partnership Project; Technical Specification Group Radio Access Network; GSM/EDGE Multiplexing and multiple access on the radio path (Release 16).

9 IEEE 802.11-2016 (2016). *Part 11: Wireless LAN Medium Access Control (MAC) and Physical Layer (PHY) Specifications*. IEEE.

10 Howitt, I. (2001). WLAN and WPAN coexistence in UL band. *IEEE Transactions on Vehicular Technology* 50 (4): 1114–1124.

11 Rappaport, T.S. (2002). *Wireless Communications: Principles and Practice*. Upper Saddle River, NJ: Prentice-Hall.

12 Lee, S.-H., Kim, H.-S., and Lee, Y.-H. (2012). Mitigation of Co-channel interference in bluetooth piconets. *IEEE Transactions on Wireless Communications* 11 (4): 1249–1254.

13 Part 15.1: Wireless Medium Access Control (MAC) and Physical Layer (PHY) Specifications for Wireless Personal Area Networks (WPANs), IEEE 802.15.1-2005, 2005.

14 IEEE Std 802.15.2-2003 (2003). IEEE Recommended Practice for Information technology-- Local and metropolitan area networks-- Specific requirements--Part 15.2: Coexistence of Wireless Personal Area Networks with Other Wireless Devices Operating in Unlicensed Frequency Bands, pp. 1–150, doi: https://doi.org/10.1109/IEEESTD.2003.94386. IEEE

15 Price, R. and Green, P.E. (1958). A communication technique for multipath channels. *Proceedings of the IRE* 46 (3): 555–570.

16 Chew, D. (2018). *The Wireless Internet of Things: A Guide to the Lower Layers*. Wiley-IEEE.

17 Burbank, J., Kasch, W., and Ward, J. (2011). *Network Modeling and Simulation for the Practicing Engineer*. Hoboken, NJ: Wiley-IEEE.

5

Secondary Spectrum Usage and Signal Detection

The underlying assumption of this book is that industry, with the help of standards providing interoperability, can use cognitive radio techniques to overcome the significant challenge of meeting increased wireless demand with finite spectrum resources. Why do we think this is possible? We think this is possible because there are many spectrum resources that are currently *underutilized*.

As described in earlier chapters, a spectrum resource is specific to a period of time, a band of frequencies (a frequency channel), and often a geographic location. A spectrum resource is said to be **underutilized** when that resource can support more traffic than is currently allocated to it. This can be measured in terms of *spectrum occupancy* (detailed later in this chapter). If a spectrum-occupancy metric indicates that a spectrum resource is being used well under capacity, then this indicates that the resources are *underutilized* and there is room for other users. Studies have shown there are regions of licensed spectrum that are underutilized. As an example, the 2002 Spectrum Policy Task Force (SPTF) of the Federal Communications Commission (FCC) [1] urged the FCC to consider *secondary use* of these resources.

Secondary Spectrum Use is a means by which primary users retain priority in accessing allocated spectrum resources, but allow other users access when idle. This necessarily dictates that secondary users yield to primary users to avoid contention. This alternative arrangement increases spectrum utilization, while maintaining the primacy of primary users.

This chapter will present two related topics, *secondary spectrum use* and *signal detection in noise*. The discussion on secondary spectrum use will begin with a definition of *spectrum occupancy* and *white space* in Section 5.1. The discussion on secondary spectrum use will continue to define Secondary Users (SUs) in Section 5.2, which concludes by detailing **Opportunistic Spectrum-Use**. The

Wireless Coexistence: Standards, Challenges, and Intelligent Solutions, First Edition.
Daniel Chew, Andrew L. Adams, and Jason Uher.
© 2021 The Institute of Electrical and Electronics Engineers, Inc.
Published 2021 by John Wiley & Sons, Inc.

chapter will then detail methods of signal detection in noise. The ability to detect the presence of signals is essential to opportunistic spectrum-use and cognitive radio in general. A unified framework for the analysis of signal detection capabilities is established in Section 5.3. The section concludes by describing the estimation of noise power and establishing decision-rule thresholds.

Also in this chapter, we explore various algorithms for signal detection employed in existing wireless coexistence standards. The first type of detector explored is the **Energy Detector** in Section 5.4. The energy detector is a blind signal detection technique found in IEEE 802.22 [2] and IEEE 802.11 [3]. The second type of signal detector this chapter will explore is the **Known Pattern Detector** (correlation) in Section 5.5. Known Pattern Detection is a *data-aided* approach found in many wireless communication standards, and is used for both signal acquisition and coarse parameter estimation. An example would be running a correlator to search for a known preamble. Lastly, this chapter will explore **Cyclic Spectral Analysis** in Section 5.6, as another means of signal detection. This method can be found in IEEE 802.22 where it is defined as *signal-specific*. That term suggests that the detector is optimized for, or limited to, one type of signal.

5.1 Spectrum Occupancy and White Space

Spectrum occupancy is a metric used to determine how much of a spectrum resource is used. From this, one can determine how much of the resource is available. Whether or not a frequency channel is *occupied* or *unoccupied* during a period of time at a specific place is based on the detection of signal activity. Signal activity is detected within the bandwidth of that frequency channel for a specified measurement period. One commonly used detection metric is *Energy*. If the energy detected in a given bandwidth exceeds a certain threshold, then that spectrum resource is said to be occupied. Energy detection is commonly employed for clear channel assessment (CCA) as described in Chapter 4.

Energy detection is not the only means to detect whether or not a spectrum resource is occupied. This Chapter will discuss three different signal detection algorithms, which all provide such an indication based on different signal *features*.

If occupancy is measured only once, then the measurement only indicates whether the spectrum resource is being used *right now*. Multiple measurements can be averaged to provide a metric of the average spectrum occupancy, determined by dividing the number of measurements resulting in an occupied status by the total number of measurements within an averaging period. This is shown in Eq. (5.1), where N_{oc} is the number of measurements that were deemed occupied

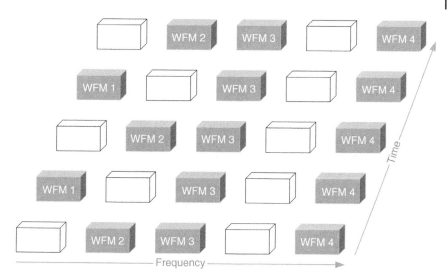

Figure 5.1 Accessing white space.

and N_{meas} is the total number of measurements. $N_{\text{meas}} = N_{\text{oc}} + N_{\text{un}}$, where N_{un} is the number of measurements deemed unoccupied.

$$\text{Average Occupancy} = \frac{N_{\text{oc}}}{N_{\text{meas}}} \tag{5.1}$$

When the spectrum is unoccupied, it is commonly referred to as **white space**. This concept is illustrated in Figure 5.1. Waveforms 1 and 2 are only active periodically, and there is an activity gap between waveforms 3 and 4. Waveforms 1 and 2 are not using their spectrum resources fully, potentially leaving room for additional wireless users. SUs are responsible for identifying these white spaces, occupying them as desired, and releasing them when primary user activity is sensed or predicted.

5.2 Secondary Users

The use of the spectrum by secondary users can be described as either *overlay* or *underlay*. The IEEE 1900.1 standard [4] provides a definition for **spectrum underlay** to include any case where a secondary user concurrently transmits on the same spectrum resource as a primary user. The underlying secondary user keeps the resulting interference within established tolerable bounds [4]. This is contrasted with **spectrum overlay**, also defined in the IEEE 1900.1 standard, in which a secondary user avoids interference with that of primary users' altogether. By this definition, Figure 5.1 illustrates an *overlay system* where the system does not attempt concurrent use of a spectrum resource but instead seeks idle resources.

This terminology is not universal in literature. A separate but related taxonomy is found in reference [5] that defines three categories of spectrum reuse: *overlay*, *underlay*, and *interweaving*. In that taxonomy, underlay specifically refers to a secondary user occupying the same time-frequency resources as a primary user, and therefore interferes with that primary user to some tolerable amount. Also, in Ref. [5], *overlay* refers to a secondary user (called the *cognitive user*) relaying the primary user's signal within the secondary user's signal, thus *overlaying* the two signals. Finally, [5] uses the term "interweaved" to describe the example shown in Figure 5.1, where the secondary user accesses white spaces between primary user activity. In doing so, the secondary user transmissions appear woven between the primary user transmissions. To be clear, "Interweaved" in Ref. [5] is what IEEE 1900.1 calls "overlaying".

This chapter will focus on *overlaying* as defined in IEEE 1900.1 and illustrated in Figure 5.1. The concept of concurrent use of spectrum resources will be addressed in more detail in Chapter 10 as *Nonorthogonal Multiple Access*.

Secondary use can be centrally coordinated to give priority to primary users, allowing secondary use only when the primary user is not present. The geolocation databases used in television white space (TVWS) are an example of a centrally coordinated system. Secondary users must check with a central repository of spectrum use to see if there is any availability.

As an alternative or in addition to centrally coordinated secondary use, SUs may be *opportunistic* meaning that the SUs monitor the spectrum and wait for opportunities to use spectrum resources. This technique necessarily employs some form of **Spectrum Sensing**. IEEE 1900.1 [4] defines Spectrum Sensing as:

a) The act of categorizing and evaluating radio signals for the purpose of obtaining information.
b) The act of measuring information indicative of spectrum occupancy (information may include frequency ranges, signal power levels, bandwidth, location information, etc.). Spectrum sensing may include determining how the sensed spectrum is used. *Source*: Adopted from IEEE 1900.4a-2011 [6]
c) The act of measuring information indicative of spectrum occupancy (information may include frequency ranges, signal power levels, bandwidth, location information, etc.) in the context of radio frequency spectrum. Sensing may include determining how the sensed spectrum is used. *Source*: Adopted from IEEE 1900.6-2011 [7]

IEEE 1900.1 elucidates that definition with the following note:

Spectrum sensing as defined in (a) pertains to the presence of signals, whereas spectrum sensing as defined in (b) and (c) pertains to sensing the presence or absence of a signal or its features for the purpose of identifying spectrum opportunities.

Sensing for the purpose of identifying spectrum *opportunities* is fundamental to **Opportunistic Spectrum-Use**. In order to maximize the use of licensed spectrum, or to avoid contention in unlicensed spectrum, a system accessing the spectrum *opportunistically* will first *sense* the spectrum for existing traffic, and only operate where and when traffic is not detected. The concept of Opportunistic Spectrum-Use has been formalized in several existing standards such as IEEE 802.22 [2] and IEEE 802.11 [3]. These standards focus on operation in *TV white space bands*, meaning unused frequency-channels found in the Television Band. The idea of opportunistically using the spectrum applies to unlicensed spectrum also. In unlicensed spectrum, there is no primary user defined, but users still have the potential to interfere with one another. If one system knew where others were operating, it could avoid them and choose spectrum locations showing no activity.

Categories of coexistence mechanisms were established in Chapter 1 and described as different *strategies*. A *sensing strategy* differs from a *mitigation strategy* in that a sensing strategy is proactive in avoiding interference. A mitigation strategy finds a way to suffer through the problem (e.g., mitigation through spread spectrum). A sensing strategy is distinct from a *monitoring strategy* as a sensing strategy does not react to interference but rather attempts to avoid interference in the first place. This makes opportunistic spectrum-use a sensing strategy and distinct from all other forms of multiple-access techniques explored in Chapter 4.

5.3 Signal Detection

Signal Detection is essential for wireless communications. A receiver must first discern if a signal is present or not before demodulation can begin. Signal Detection is of particular importance to opportunistic spectrum-use. In such a scheme, the SU must detect white space in the spectrum to discern if opportunities for spectrum use exist.

Signal Detection allows us to identify the presence of signal activity with respect to both time and frequency. From this, we can begin to map out the spectrum in order to make an informed decision about where to operate to avoid contention. The algorithms used for signal detection depend in part on our level of prior information, i.e., *"what do we know about the signal?"* With some prior knowledge, we can choose algorithms and search strategies specific to the expected signal activity. These types of algorithms are called **data-aided** or **signal-specific**. Data-aided and signal-specific are widely used synonymously, but the two names imply a difference. Data-aided detection is typically done for a known data sequence, such as a preamble defined in a wireless standard. Quite literally, a data-aided detection method requires that the receiver know the exact transmitted data in advance. Signal-specific detectors are a broader class. A detector may be signal-specific but may also not know, precisely, what the transmitted data will be. For example, a signal-specific detector may focus on the fact that the signal to be detected contains a pilot tone, and the signal-specific detector can place a narrow filter at the expected location of that pilot tone. A signal-specific search yields improved performance with respect to noise, but lacks the generality to handle signals previously unseen.

If little or no information is known about the signal to be detected beforehand, then we execute a **blind search**, and look for any signal activity. A blind search is typically more susceptible to errors due to noise than a signal-specific method, but has the generality necessary to handle signals previously unseen as compared to the signal-specific alternatives.

One key question is *"how blind is blind?"* Where, precisely, is the line between these different categories? There is no standard answer to this question. To say that a sensing technique is *blind* is to imply that it requires no assumptions on the type of signal to be sensed. The IEEE 802.22 standard offers some disambiguation by adding a second dimension to the categorization. In that standard, sensing techniques are categorized as either *Blind* or *Signal-Specific* and also as either **Coarse** or **Fine**. Coarse methods are intended to run for short measurement intervals. Fine methods are intended to guarantee detection of a specific signal at a specific strength, but may be general enough to also detect other signals. This will be discussed in greater detail in Chapter 8.

The terms *signal-specific* and *blind* give an indication as to which *features* are used to make decisions about the presence of signal activity. *Feature selection* for signal detection is discussed in Section 5.3.3. However, before a discussion on feature selection can begin, we must establish a mathematical framework and that begins with a definition of *Binary Hypothesis Testing*.

5.3.1 Binary Hypothesis Testing

Signal Detection is an example of a **Binary Hypothesis Test**. A Binary Hypothesis Test attempts to discern which of two possible hypotheses is true. Hypothesis testing is covered in detail in Refs. [8] and [9], but because Binary Hypothesis Testing is fundamental to Signal Detection, this section will provide a brief overview.

The binary hypothesis test is conducted using observed data $\vec{r}$. An **observation** of a random variable or vector is the value that has been measured in some instance. This vector $\vec{r}$ will be referred to as *raw measurements*. The raw measurements (observations) $\vec{r}$ come from a random vector $\vec{R}$, which follows an unknown distribution. There are two competing hypotheses to explain and model $\vec{R}$. The two hypotheses are called the *null hypothesis* (H_0) and the *alternative hypothesis* (H_1). The two hypotheses are shown in Eq. (5.2). If H_0 is true, then $\vec{R}$ will follow probability distribution $\mathcal{R}_0$. If H_1 is true, then $\vec{R}$ will follow probability distribution $\mathcal{R}_1$.

$$H_0 : \vec{R} \sim \mathcal{R}_0$$

$$H_1 : \vec{R} \sim \mathcal{R}_1 \tag{5.2}$$

A **decision metric** is used to make the determination between the two hypotheses. This decision metric is calculated $\vec{r}$ by way of function $d(\vec{r})$. A decision rule is then applied to the decision metric, as shown in the example in Eq. (5.3). In this decision rule, the decision metric is compared to a threshold η. If the decision metric $d(\vec{r})$ is greater than η, then H_1 is chosen; if $d(\vec{r})$ is less than or equal to η, H_0 is chosen.

$$d(\vec{r}) \underset{H_1}{\overset{H_0}{\lessgtr}} \eta \tag{5.3}$$

This decision between H_0 and H_1 leads to two different types of possible errors, as shown in Table 5.1. The rows indicate which of the hypothesis was actually true. The columns indicate which of the two hypotheses was chosen. If the truth is chosen, the state of Table 5.1 is simply *Correct*. However, if the truth is not chosen, then there are two distinct errors. A Type I error is a *false positive*, meaning that H_1 was chosen but H_0 was true. A Type II error is a *false negative*, meaning that H_0 was chosen but H_1 was true.

Signal Detection selects between the two hypotheses shown in Eq. (5.4) where $r[k]$ represents the measurements in time at the receiver if every element in $\vec{r}$ were given a sample index $[k]$. The variable $r[k]$ may be composed only of noise in the receiver bandwidth. The component $s[k]$ represents some signal, other than noise,

Table 5.1 Binary hypothesis testing results.

	Choose H_0	Choose H_1
H_0 True	Correct	Type I error
H_1 True	Type II error	Correct

Table 5.2 Signal detection testing results.

	Choose H_0	Choose H_1
H_0 True	No Activity	False Alarm
H_1 True	Missed Detect	Signal Detected

which may or may not be present in the receiver bandwidth. The hypothesis test attempts to determine if such a signal is present. If H_0 is true then there is no signal present in the bandwidth under examination. If H_1 is true then a signal is present in the bandwidth under examination.

$$H_0 : r[k] = n[k]$$

$$H_1 : r[k] = s[k] + n[k] \tag{5.4}$$

Just like the more generic case, there are two error categories in signal detection. Those two error categories are **False Alarm** and **Missed Detect**. A False Alarm (FA), which is a false positive, occurs when the detector erroneously detects a signal as present. A Missed Detect (MD), which is a false negative, occurs when the detector fails to detect a signal when present. Those errors are mapped in Table 5.2, which follows the same paradigm as Table 5.1, but uses terminology common to communication systems and radar.

5.3.2 A Generic Framework for Signal Detection

Many different types of signal detectors are used in practice to determine if a spectrum resource is in use. The IEEE 802.22 standard identifies over a dozen techniques, each with different strengths and weaknesses. The motivation behind this generic framework for signal detection is to provide the reader a unified approach to signal detection. In order to provide a thoughtful discussion on signal detection, a common basis for comparison between varied techniques must be established.

The role of signal detection is to determine if a signal is present on a frequency channel at a given time and location. The process is illustrated in Figure 5.2, and begins with samples collected from the received signal, which are in the raw data vector $\vec{r}$. The next stage in the process extracts *features* from the raw data. A **feature** of the signal is a quantitative measurement of some attribute of that signal. Signal features include but are not limited to: *energy, bandwidth, center frequency, duration in time, and symbol rate.* Careful selection of these features is essential to the success of signal detection. The selection of features for signal detection is discussed in Section 5.3.3. These features are extracted from $\vec{r}$ by

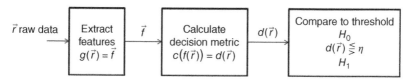

Figure 5.2 The signal detection process.

function $g(\vec{r})$, which yields the vector of features $\vec{f}$. The function $g(\vec{r})$ may extract multiple features from the raw data $\vec{r}$, using both linear and non-linear processes. A second function $c(\vec{f})$ is used to calculate a decision metric from the feature vector. The decision metric is a single value calculated from these features. In order to avoid confusion, two functions are shown, $c(\vec{f})$ and $d(\vec{r})$. The function $d(\vec{r})$ is a simplification of the fact that the calculation of a decision metric from feature vector $\vec{f}$ is ultimately a function of the raw data $\vec{r}$. This is to say: $c(\vec{f}) = c(g(\vec{r})) = d(\vec{r})$. Finally, the decision metric $d(\vec{r})$ is used to choose between the two hypotheses by comparing it to threshold η.

The process illustrated in Figure 5.2 is a generic framework for signal detection. All signal detection methods follow this paradigm, whether looking for a sequence of symbols, or simply detecting the presence of energy.

Any instance of this framework must define three functions:

- The function $g(\vec{r})$ which extracts features from the raw data.
- The function $c(\vec{f})$ which determines the decision metric, and
- The value of threshold η

There are many different types of detectors, each with different definitions of these three functions. The details of defining these functions can be complex, requiring the study of multiple subjects. The following subsections provide some details in how these functions might be defined, starting with feature selection and ending with a discussion on *Constant False Alarm Rate* (CFAR).

5.3.3 Feature Selection

The goal of feature selection is to choose the extracted features such that the signal detector for a particular application may enjoy low rates of error. This process requires an analysis of how the features contribute to the decision metric. This process also requires that one consider what features can reliably be extracted given a set of target signals.

There are many considerations in feature selection for signal detection, which requires significant domain knowledge in wireless communications. One example is whether the receiver has some knowledge of the signal to be detected. Examples include pilot tones and specific data sequences. If such known parameters exist then those signal-specific parameters can be estimated and those estimates become features in the feature set. Such features are used in *signal-specific* detectors, and having such knowledge improves detection probabilities. Now consider the case where there are many possible types of signals without consistent parameters. In such a case, the signal detector must rely on features that are not dependent upon the idiosyncrasies of a specific signal. Simply because a given feature is useful for detecting one type of signal does not necessarily mean it is useful in detecting others. For a signal detector to be nonsignal-specific, features must be *generalizable* across many different and potentially unknown signals. These types of features are used in *blind* detectors.

The features are the *evidence* that will be used to choose between the two hypotheses. The feature vector $\vec{f}$ is derived from a random process. Therefore, the probability distribution of the features must provide some means to distinguish between these hypotheses. As such, selecting features that can serve this task involves modeling their distribution.

The conditional probability $P(\vec{f} \mid H_k)$ is the probability of observing a specific vector of features $\vec{f}$ given hypothesis H_k. This conditional probability $P(\vec{f} \mid H_k)$ is also called the **Likelihood**. For signal detection, Likelihood is the conditional probability of the evidence given one of two hypotheses, *signal present* or *signal not present*. Those probabilities are $P(\vec{f} \mid H_1)$ and $P(\vec{f} \mid H_0)$, respectively.

The Likelihood will now be used to define the Maximum Likelihood Detector. The probability of evidence will be revisited in Section 5.3.5.

5.3.4 Maximum Likelihood Detector

If we design the signal detector using the **Maximum Likelihood Rule**, then we have designed the detector to choose between H_0 and H_1 based on which hypothesis yields the greater *Likelihood*. This is expressed in Eq. (5.5). If the probability of observing a vector of features $\vec{f}$ given H_0 is greater than or equal to the probability of observing a vector of features $\vec{f}$ given H_1, then choose H_0. However, if the probability of observing a vector of features $\vec{f}$ given H_1 is greater than the probability of observing a vector of features $\vec{f}$ given H_0, then choose H_1.

$$P(\vec{f} \mid H_1) \underset{H_1}{\overset{H_0}{\lessgtr}} P(\vec{f} \mid H_0) \tag{5.5}$$

The two Likelihoods can be combined into a ratio, as shown in Eq. (5.6). If this ratio is less than 1, choose H_0. If the ratio is greater than 1, choose H_1. This is called the **Likelihood Ratio**.

$$\frac{P\left(\vec{f} \mid H_1\right)}{P\left(\vec{f} \mid H_0\right)} \underset{H_1}{\overset{H_0}{\lessgtr}} 1 \tag{5.6}$$

From Eq. (5.6), we can now define a function $c_{LR}(\vec{f})$ to calculate the decision metric from the feature vector $\vec{f}$. This is shown in Eq. (5.7). The subscript LR in $c_{LR}(\vec{f})$ indicates this is the *Likelihood Ratio*.

$$d_{LR}\left(\vec{r}\right) = c_{LR}\left(\vec{f}\right) = \frac{P\left(\vec{f} \mid H_1\right)}{P\left(\vec{f} \mid H_0\right)} \tag{5.7}$$

Also from Eq. (5.6), we see that setting the threshold η to 1 produces the *Maximum Likelihood decision rule* shown in Eq. (5.8).

$$d_{LR}\left(\vec{r}\right) \underset{H_1}{\overset{H_0}{\lessgtr}} 1 \tag{5.8}$$

It is often more efficient to calculate the **Log-Likelihood Ratio** as derived in Eqs. (5.9) and (5.10). This converts a computationally costly or unstable division operation into subtraction.

$$\log\left(\frac{P\left(\vec{f} \mid H_1\right)}{P\left(\vec{f} \mid H_0\right)}\right) \underset{H_1}{\overset{H_0}{\lessgtr}} \log(1) \tag{5.9}$$

$$\log\left(P\left(\vec{f} \mid H_1\right)\right) - \log\left(P\left(\vec{f} \mid H_0\right)\right) \underset{H_1}{\overset{H_0}{\lessgtr}} 0 \tag{5.10}$$

When using log-likelihoods, the decision rule becomes Eq. (5.11), where the threshold η is set to 0 and the subscript "LLR" indicates the *Log-Likelihood Ratio*.

$$d_{LLR}\left(\vec{r}\right) \underset{H_1}{\overset{H_0}{\lessgtr}} 0 \tag{5.11}$$

In this section, we have established the *Maximum Likelihood decision rule*. However, the derivation presented in this section is missing a crucial detail. The Maximum Likelihood decision rule depends on being able to compute $P(\vec{f} \mid H_k)$ for both H_1 and H_0.

For the case of H_0, the features are drawn from Additive White Gaussian Noise (AWGN). If we assume the noise is zero mean, we need only estimate the variance to compute $P(\vec{f} \,|\, H_0)$.

However, defining $P(\vec{f} \,|\, H_1)$ is a more complex problem. In order to define $P(\vec{f} \,|\, H_1)$, which is the conditional probability of $\vec{f}$ given that a signal is present, we need to be able to reliably model the feature vector $\vec{f}$ for the signal or signals we intend to detect. It may be difficult to select a set of features for which $P(\vec{f} \,|\, H_1)$ can be defined over a wide variety of signals. Recall that selecting features which can be extracted from a wide variety of signals is necessary for *blind* detection. If successful, it can be said that such a detector *generalizes* well across numerous applications. If the types of signals being detected are limited, or even just one type of signal, then the signals can be modeled with greater specificity. This greater specificity allows the selection of *signal-specific* features to be used in defining $P(\vec{f} \,|\, H_1)$.

5.3.5 Maximum A Posteriori (MAP) Detector

An improvement over selecting the greater likelihood can be found in selecting the greater **posterior probability**. The posterior probability $P(H_k \,|\, \vec{f})$ is the conditional probability of a hypothesis being true given a set of feature values. It is called the *posterior* probability because it is the probability of a hypothesis being true *after* the features have been measured. The Maximum A Posteriori (MAP) detector selects the hypothesis with the greater probability of being true given a feature vector $\vec{f}$. The MAP decision rule is shown in Eq. (5.12).

$$P\!\left(H_1 \,|\, \vec{f}\right) \underset{H_1}{\overset{H_0}{\lessgtr}} P\!\left(H_0 \,|\, \vec{f}\right) \tag{5.12}$$

We can use Bayes Theorem to change the terms in Eq. (5.12) into Eq. (5.13). In Eq. (5.13), we use the Likelihood, $P(\vec{f} \,|\, H_k)$, and *prior probability*, $P(H_k)$, in the MAP decision rule.

$$P\!\left(\vec{f} \,|\, H_1\right)P(H_1) \underset{H_1}{\overset{H_0}{\lessgtr}} P\!\left(\vec{f} \,|\, H_0\right)P(H_0) \tag{5.13}$$

Moving the terms on both sides of the decision rule yields Eq. (5.14).

$$\frac{P\!\left(\vec{f} \,|\, H_1\right)}{P\!\left(\vec{f} \,|\, H_0\right)} \underset{H_1}{\overset{H_0}{\lessgtr}} \frac{P(H_0)}{P(H_1)} \tag{5.14}$$

We define the decision metric for MAP as was done for ML in Eq. (5.7) as is shown in Eq. (5.15)

$$d_{MAP}\left(\vec{r}\right) = c_{MAP}\left(\vec{f}\right) = \frac{P\left(\vec{f} \mid H_1\right)}{P\left(\vec{f} \mid H_0\right)} \tag{5.15}$$

The MAP decision rule can be expressed in terms of the decision metric $d_{MAP}(\vec{r})$, as shown in Eq. (5.16).

$$d_{MAP}\left(\vec{r}\right) \underset{H_1}{\overset{H_0}{\lessgtr}} \frac{P(H_0)}{P(H_1)} \tag{5.16}$$

$P(H_k)$ is the probability of a specific hypothesis H_k being true. The probability $P(H_k)$ is called a **prior probability**. *Consider an example*: There is a frequency channel where a primary user signal may be present at random. Let us say, we know that for any occupancy measurement, there is a 25% chance of the signal being present. This means that the probability of a signal being present is 25% and the probability of a signal not being present is 75%. Therefore, the probability $P(H_0) = 0.75$ and $P(H_1) = 0.25$. In this scenario, we know this information *prior to any evidence*, hence the term *prior probability*.

If the prior probabilities are known, then they can be used to optimize the decision threshold. This provides for an improvement over the maximum likelihood detector. However, in many cases, we cannot know the prior probabilities, making the MAP detector unrealizable.

5.3.6 Probability of Error

Recall that the decision metric, $c(\vec{f}) = d(\vec{r})$, is calculated from the feature vector. The feature vector is random, therefore the decision metric is also random. This randomness provides a chance for an error. Any one instance of observed values of the features may yield a decision metric that would lead to an erroneous decision being made.

We can describe the probability of $d(\vec{r})$ given a hypothesis, $P(d(\vec{r}) \mid H_k)$ much as was done for the feature vector. This will be important in determining the probability of error.

As first described in Section 5.3.1, the outcomes of the decision metric test are now shown in Table 5.3 accompanied by the test criteria. In each of the four cells there is now a cumulative conditional probability.

Consider the Maximum Likelihood and MAP decision rules. Each rule chooses the greater probability, but it is still a probability and that probability is not 100%.

Table 5.3 Signal detection testing probabilities.

	$d(\vec{r}) \leq \eta$ **Choose** H_0	$d(\vec{r}) > \eta$ **Choose** H_1
H_0 True	$P(d(\vec{r}) \leq \eta \mid H_0)$No Activity	$P(d(\vec{r}) > \eta \mid H_0)$False Alarm
H_1 True	$P(d(\vec{r}) \leq \eta \mid H_1)$Missed Detect	$P(d(\vec{r}) > \eta \mid H_1)$Signal Detected

The decision rules are designed to choose the *most likely hypothesis*, but not necessarily the *correct hypothesis*. As alluded to earlier, there is always a possibility that $d(\vec{r})$ could be found greater than the threshold η with no signal present (*False Alarm*). There is also a possibility that $d(\vec{r})$ could be found lesser than the threshold η with a signal present (*Missed Detect*).

The cumulative conditional probability $P(d(\vec{r}) \leq \eta \mid H_1)$ is the probability that the decision metric test did not detect signal activity; however, there was activity present. This is an *incorrect decision* and it is the **Probability of Missed Detect**, $P_{MD}(d(\vec{r}))$, shown in Eq. (5.16).

The complementary cumulative conditional probability $P(d(\vec{r}) > \eta \mid H_0)$ is the probability that the decision metric test detected signal activity; however, there was no activity present. This is an *incorrect decision* and it is the **Probability of False Alarm**, $P_{FA}(d(\vec{r}))$, shown in Eq. (5.17).

$$P_{MD}(d(\vec{r})) = P(d(\vec{r}) \leq \eta \mid H_1) \tag{5.16}$$

$$P_{FA}(d(\vec{r})) = P(d(\vec{r}) > \eta \mid H_0) \tag{5.17}$$

To elaborate further, consider the example illustrated in Figure 5.3. In this example, the conditional probabilities $P(d(\vec{r}) \mid H_0)$ and $P(d(\vec{r}) \mid H_1)$ both follow Gaussian distributions. In this example, $d(\vec{r}) \mid H_0$ follows the zero-mean distribution $\mathcal{N}(0, \sigma_n^2)$ and $d(\vec{r}) \mid H_1$ follows $\mathcal{N}(A, \sigma_n^2)$, where A is the amplitude of the received signal. The probability of false alarm is the area under the curve of $P(d(\vec{r}) \mid H_0)$ where $d(\vec{r}) > \eta$. The probability of missed detect is the area under the curve of $P(d(\vec{r}) \mid H_1)$ where $d(\vec{r}) \leq \eta$. These identified areas both grow or shrink depending on the value of η.

5.3.7 Choosing the Threshold for a False Alarm Rate

It may be the case that the signal detector cannot know how often a signal will be present or at what power level that signal will be received. What the signal detector can know is the power of the noise within its own receiver bandwidth. We know that this has a Gaussian (Normal) distribution (see Chapter 3) with a mean of zero

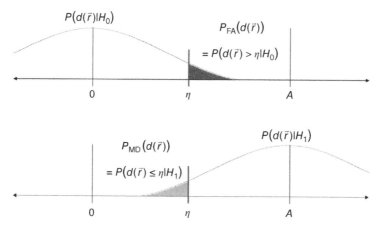

Figure 5.3 Probability of false alarm and missed detect.

and some variance σ_n^2. We do not yet know the value of the noise variance. In Section 5.3.9, we will establish means to estimate the noise variance. For now, let us assume, we have that estimator working and we can accurately estimate the noise variance.

The *probability of false alarm*, $P_{FA}(d(\vec{r}))$, is a function of both the noise power in the receiver bandwidth and the threshold value. As such, a threshold can be chosen to provide a constant probability of a false alarm, with no bearing on the probability of missed detection. The probability of false alarm can be established by setting $P(d(\vec{r}) \geq \eta_{FA} \mid H_0)$ equal to a constant and then solving for η_{FA}. This requires that the cumulative conditional probability distribution for the decision metric given H_0 be known.

5.3.8 Choosing Threshold for a Missed Detect Rate

A wireless standard may specify that a certain type of detector meet a minimum probability of detection given a signal with a specified received power. In such a case, the threshold must be chosen to meet this requirement. An example of this would be in IEEE 802.11 where a frequency channel is determined as *Busy* (*Signal Present*) with greater than 90% probability of detection, if any signal with an average power of −62 dBm (decibels relative to milliwatts) is present over a 4 μs (microsecond) measurement period. Given that power is the derivative of energy with respect to time, these parameters let the system designer know exactly how much energy will be present over that 4 μs period. Based on that value of energy, a threshold can be selected to meet these criteria.

Note how very specific things need to be known in order to specify a threshold to meet a desired probability of missed detect. Unlike the probability of false alarm, the probability of missed detect depends on both noise and a specific signal model.

The probability of missed detect can be established by setting $P(d(\vec{r}) < \eta_{\mathrm{MD}} \mid H_1)$ equal to a constant and then solving for η_{MD}. However, one must be able to model $P(d(\vec{r}) < \eta_{\mathrm{MD}} \mid H_1)$. In the 802.11 energy detector example, the feature specified is energy, which generalizes well because all signals with a specified average power deliver a specific amount of energy within a given measurement period. What is not known is how many samples the implementation will use in that period of time, and what the noise variance σ_n^2 will be.

5.3.9 Noise Power Estimation

In order to set the threshold to an acceptable probability of error, the detector must estimate the noise variance σ_n^2. There are a number of techniques for receiver noise power estimation. Some of the simplest receiver noise power estimation techniques require quiet periods during which receiver noise power is directly measured. Other techniques can estimate noise power in the presence of signals. Because we typically cannot assume there are inactive channels or portions of the spectrum we can use to estimate noise power, we submit that noise power estimation in the presence of signal activity is fundamental to the cognitive functionality discussed in this text. Here, we describe the noise power estimation technique discussed in Refs. [10–12]. This noise power estimation technique is based upon the principle that if the signal and noise are independent, the received sample covariance matrix, Σ_y, can be decomposed into its signal and noise components. This is shown in Eq. (5.18), where Σ_s and Σ_n are the respective signal and noise components, and I_L is an Lth-order identity matrix.

$$\Sigma_y = \Sigma_s + \Sigma_n = \Sigma_s + \sigma_n^2 I_L \tag{5.18}$$

Once separated, we can then analyze only the components representing noise to estimate the noise power σ_n^2.

We start by estimating the sample covariance matrix from a finite block of received samples. Recall that we defined these as $\vec{r}$ earlier in this chapter. Assuming $\vec{r}$ to contain at least L N-sample observations, we can construct an $N \times L$ matrix Y, and estimate the sample covariance matrix as Eq. (5.19). Each observation $\vec{r_l}$ becomes a column in the matrix Y, where the subscript l indicates an index in time. This method necessarily assumes a greater number of observations than the length of each individual observation; that is, $L > N$. The vector $\vec{\mu}_Y$ is an L

length column-vector containing the means of each observation $\vec{r_i}$. The matrix Y^H is the conjugate transpose of Y, and the vector $\vec{\mu}_Y{}^H$ is the conjugate transpose of $\vec{\mu}_Y$.

$$\hat{\Sigma}_y = \frac{YY^H}{L-1} - \vec{\mu}_Y\vec{\mu}_Y{}^H \tag{5.19}$$

The next step is to find the N eigenvalues of $\hat{\Sigma}_y$, which we annotate as λ. If sorted in descending order, the first K eigenvalues represent signal activity and the $K+1$ thru N eigenvalues represent noise.

To estimate K, we use the *Minimum Descriptive Length* (MDL) criterion [13], which suggests we choose the minimum number of parameters to describe a particular model. As such, we iterate over the sorted (descending) eigenvalues per Eq. (5.20), where $\varphi(k)$ represents the geometric mean Eq. (5.21), and $\theta(k)$ the arithmetic mean Eq. (5.22) of the first $N-k$ sorted eigenvalues. The index representing the minimum value is then our estimate of K.

$$K = \operatorname*{argmin}_{0 \leq k < N}\left(-L(N-k)\log\left(\frac{\varphi(k)}{\theta(k)}\right)\frac{1}{2}k(2N-k)\log L \right) \tag{5.20}$$

$$\varphi(k) = \prod_{j=k}^{N-1} \lambda_j^{\frac{1}{N-k}} \tag{5.21}$$

$$\theta(k) = \frac{1}{N-k}\sum_{j=k}^{N-1} \lambda_j \tag{5.22}$$

Now having estimated K, we can estimate the noise power $\hat{\sigma}_n{}^2$. The $K+1$ thru N eigenvalues weakly follow a *Marchenko–Pastur* distribution. We can use any number of curve fitting methods to estimate the Marchenko–Pastur probability-distribution function described in Eqs. (5.23)–(5.25). The resultant standard deviation $\hat{\sigma}_n$ is then used to estimate the noise power as $\hat{\sigma}_n{}^2$. Finally, given our estimate of the noise floor, we can use this to set the detection threshold per our desired P_{FA} or P_{MD}.

$$P(x) = \frac{\sqrt{(x-a)(b-x)}}{2\pi\hat{\sigma}_n{}^2\left(\frac{L}{N}\right)}, \text{ for } a \leq x \leq b \tag{5.23}$$

$$a = \hat{\sigma}_n^2\left(1 - \sqrt{\frac{L}{N}}\right)^2 \tag{5.24}$$

$$b = \hat{\sigma}_n^2\left(1 + \sqrt{\frac{L}{N}}\right)^2 \tag{5.25}$$

5.4 Energy Detector

The **Energy** of a signal is proportional to the square of the magnitude of the received samples. The total energy of the raw data $\vec{r}$ is calculated in Eq. (5.26), where each sample $\vec{r}[k]$ is an element of $\vec{r}$.

$$E_r = \sum_{k=0}^{K-1} \left| \vec{r}[k] \right|^2 \tag{5.26}$$

Energy can be calculated from any signal, and is not dependent on any synchronization. These aspects make Energy a popular feature to extract for blind signal detection.

The Energy Detector [14], called a *radiometer* in some literature, consists of a **Non-Coherent** summation of the received signal over some number of samples. The summation is noncoherent because the phases of the samples are disregarded, and the output is a function of the magnitude alone.

5.4.1 Single-Channel Operation

The single-channel energy detector is illustrated in Figure 5.4. The raw data are a time series of complex-valued baseband samples, therefore $\vec{r}$ is a complex-valued vector with K elements. The magnitude of these complex-valued elements are calculated and then squared. The resulting squared-magnitude is compared to a threshold, and a hypothesis is chosen based on that result.

If no signal is present, then the received signal consists solely of zero-mean Gaussian noise. This follows a *scaled chi-square distribution* with $2K$ degrees of freedom. There are two components to each sample, in-phase (real) and quadrature phase (imaginary). Therefore, each sample adds two degrees of freedom. The scaling is a result of the fact that the noise may not have unit variance ($\sigma_n^2 = 1$). If the noise in the receiver bandwidth is scaled such that the noise has unit variance, then the measurement of energy given no signal is present will follow the traditional *chi-square distribution* with $2K$ degrees of freedom. The noise variance can be normalized by way of applying linear scaling to the received samples.

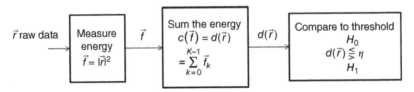

Figure 5.4 The single-channel energy detector.

The measurement of energy given a signal is present follows a *scaled noncentral chi-square distribution* with 2K degrees of freedom. The distribution is *scaled* because the noise is not guaranteed to have unit variance. The received signal could be scaled such that the noise component has unit variance, in which case, the distribution becomes the traditional *noncentral chi-square distribution* with 2K degrees of freedom. If the noise is scaled to have a unit variance, it is important to note that the scaling will also apply to any signal component in the received samples. This is because the scaling is linear and therefore superposition applies.

The probability of a missed detection and probability of a false alarm are not symmetric for the energy detector. The probability of a missed detection is given in Eq. (5.27) with the following substitutions:

- σ_n is the standard deviation of the noise
- K is the number of complex-valued samples measured
- η is the threshold
- E_s is the total energy of the signal component ($\vec{s}$) of $\vec{r}$ as measured over K complex-valued samples (*without noise*)
- Q_K is the Marcum-Q function.

The probability of false alarm is given in Eq. (5.28) where $\Gamma(\bullet)$ is the gamma function and $\Gamma(\bullet, \bullet)$ is the normalized upper incomplete gamma function. Proofs for Eqs. (5.27) and (5.28) are provided in Ref. [15].

$$P_{MD} = 1 - Q_K\left(\frac{\sqrt{E_s}}{\sigma_n}, \frac{\sqrt{\eta}}{\sigma_n}\right) \tag{5.27}$$

$$P_{FA} = \Gamma\left(K, \frac{\eta}{2\sigma_n{}^2}\right) \tag{5.28}$$

The gamma function is defined in Eq. (5.29). The upper incomplete gamma function is defined in Eq. (5.30). The Marcum-Q function is defined in Eq. (5.31). The function $I_{K-1}(\bullet)$ is the modified Bessel function of the first kind. Most Digital Signal Processing simulation tools have some implementation of these functions. The reader is advised to check if the simulation tool has implemented the functions as defined here. Note that, the incomplete gamma function is not necessarily *normalized* which means the reader would need to implement the normalization $\frac{1}{\Gamma(K)}$.

$$\Gamma(K) = (K-1)! \tag{5.29}$$

$$\Gamma(K,b) = \frac{1}{\Gamma(K)}\int_b^\infty x^{K-1}e^{-x}dx \tag{5.30}$$

$$Q_K(a,b) = \left(\frac{1}{a}\right)^{K-1} \int\limits_b^\infty x^K e^{-\frac{(x^2+a^2)}{2}} I_{K-1}(ax)dx \qquad (5.31)$$

Reference [15] builds upon the signal model adding fading as a random parameter affecting the received signal power. For the purposes of this discussion, Eqs. (5.27) and (5.28) are sufficient and assume the energy of the signal over the measurement period is constant. Equations (5.27) and (5.28) provide the basis for P_{MD} and P_{FA} in the case of noncohrent summation.

The probabilities for P_{MD} and P_{FA} are illustrated in Figure 5.5. The noise in the receiver bandwidth was set to unit variance. The number of samples was set at 80 due to the fact that this yields an observation period of 4 µs at 20 million samples per second (MSPS) which makes this relevant to the energy detector described in the IEEE 802.11 standard. The x-axis is the value of the threshold, and the y-axis is probability. The probability of a missed detect and the probability of a false alarm are plotted for three different received Signal-to-Noise power Ratios (SNR), 0, 1, and 2 dB. This illustration shows that the probability of a false alarm does not

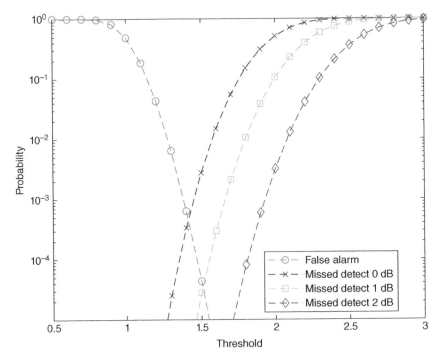

Figure 5.5 Energy detector operation given different SNR values.

change as a function of SNR. This is because the noise power in the receiver bandwidth is constant. Because the received signal power varies in the three cases, the probability of a missed detect changes as a function of SNR. A tolerable probability of false alarm can be selected on the y-axis, and then a corresponding value on the x-axis can be selected as the threshold.

Solving for a threshold for a desired probability of false alarm using Eq. (5.28) is shown in Eq. (5.32) where $\Gamma^{-1}(\bullet, \bullet)$ is the inverse upper incomplete gamma function. This function is available in simulation tools such as *MATLAB*.

$$\eta_{\text{FA}} = 2\sigma_n{}^2 \Gamma^{-1}(K, P_{\text{FA}}) \tag{5.32}$$

Solving for a threshold using Eq. (5.27) can be difficult as that requires the inverse of the Marcum Q function. As an alternative to finding the inverse of the Marcum Q function of some K samples, an approximation can be used. One such approximation is called the Berkeley Model, and is described in Ref. [16]. In that model, a large K is used and the decision metric is approximated as Gaussian. The Gaussian distribution for H_0 has a mean of $2K\sigma_n{}^2$ and a variance of $4K\sigma_n{}^4$. The Gaussian distribution for H_1 has a mean of $2K(\sigma_s{}^2 + \sigma_n{}^2)$ and a variance of $4K(\sigma_s{}^2 + \sigma_n{}^2)^2$ where $\sigma_s{}^2$ is the variance of the signal that is present. This is the same technique discussed in the IEEE 802.22 standard, except there the power of noise is taken as the sum of the variance in the real and imaginary components, $2K\sigma_n{}^2$. The approximated probability of a missed detect, $P_{\widetilde{\text{MD}}}$, is shown in Eq. (5.33) and the approximated probability of false alarm, $P_{\widetilde{\text{FA}}}$, is shown in Eq. (5.34). Both of these expressions are based on the complementary cumulative distribution of a Gaussian distribution represented by the Q function, which is defined in Eq. (5.35).

$$P_{\widetilde{\text{MD}}} = Q\left(\frac{\eta - 2K(\sigma_s{}^2 + \sigma_n{}^2)}{(\sigma_s{}^2 + \sigma_n{}^2)2\sqrt{K}}\right) \tag{5.33}$$

$$P_{\widetilde{\text{FA}}} = Q\left(\frac{\eta - 2K\sigma_n{}^2}{2\sigma_n{}^2\sqrt{K}}\right) \tag{5.34}$$

$$Q(a) = \frac{1}{\sqrt{2\pi}} \int_a^\infty e^{\frac{-x^2}{2}} dx \tag{5.35}$$

The threshold for a missed detection is shown in Eq. (5.36) using the Berkeley Model. $Q^{-1}(\bullet)$ is the inverse Q function and it is available in tools such as *MATLAB*. Note: *in the original Ref.* [16], *the number of samples counts the real and imaginary components as separate contributors.*

$$\eta_{\widetilde{\text{MD}}} = 2(\sigma_s{}^2 + \sigma_n{}^2)\left(Q^{-1}\left(P_{\widetilde{\text{MD}}}\right)\sqrt{K} + K\right) \tag{5.36}$$

5.4.2 Multichannel Operation

Concurrent energy detectors can be run on multiple frequency channels, enabling a secondary user to detect white space over a range of potential frequency channels.

Suppose that $\vec{r}$ represents a large bandwidth. This wide bandwidth can be channelized into multiple, individual, frequency-channels. Individual energy detectors can then be run on the resulting frequency channels.

Consider an example of a multichannel operation using orthogonal frequencies. The samples of the vector $\vec{r}$ will be frequency downconverted by being mixed with the conjugate of a complex sinusoid centered at $\frac{2\pi m}{M}$, where m is the index of the feature being extracted and M is the total number of features. This is shown in Eq. (5.37). The vector of received samples $\vec{r}$ will be of length K.

$$\vec{f}[m] = \sum_{k=0}^{K-1} \vec{r}[k] e^{-i\frac{2\pi m}{M}k} \tag{5.37}$$

The multichannel energy detector can employ the matrix operation shown in Eq. (5.38) where $\mathbf{F}$ is a matrix of discrete time Fourier basis functions.

$$\vec{f} = \mathbf{F}\vec{r} \tag{5.38}$$

The Fast Fourier Transform (FFT) can be used if $M = K$ and K is set to a power of two. Each extracted feature corresponds to one frequency bin of the FFT. Note that the summation in each frequency bin is a *coherent summation*. After computing all M features, the multichannel detector will calculate M decision metrics, as shown in Eq. (5.39).

$$c\left(\vec{f}[m]\right) = \left|\vec{f}[m]\right|^2 \tag{5.39}$$

Now each decision metric can be compared to the threshold η. The probability of a missed detect in bin m is a function of the signal energy present in bin m, denoted as E_m. The probability of false alarm is the same in all bins, assuming a flat noise floor derived from AWGN. $Q_1(\bullet, \bullet)$ is the first-order Marcum Q function. It follows the definition in Eq. (5.31).

$$P_{MD}[m] = 1 - Q_1\left(\frac{\sqrt{E_m}}{\sqrt{M}\sigma_n}, \frac{\sqrt{\eta}}{\sqrt{M}\sigma_n}\right) \tag{5.40}$$

$$P_{FA} = e^{\frac{-\eta}{2M\sigma_n^2}} \tag{5.41}$$

5.5 Known Pattern Detector

It is commonplace for wireless signals to contain at least one predetermined pattern. These patterns are often defined by the standard for a variety of purposes, such as synchronization and channel estimation. A common example is a **preamble** (*Discussed in Chapter* 3). A preamble is a specific sequence of values appended to the start of a transmission. The receiver uses this knowledge to perform cross-correlation on its received samples. If the correlation value exceeds a chosen threshold, the desired signal is assumed present, and the cross-correlation values can also be used to estimate the start of the framing of data and other useful parameters. In performing this calculation, the receiver is detecting the presence of a *known pattern* in the received waveform.

5.5.1 Calculation in the Time-Domain

Detection of a known sequence can be done by way of correlation and is shown in Eq. (5.42). The samples of the known sequence are stored in a vector $\vec{w}$ of length L. The known sequence $\vec{w}$ is then cross-correlated element-by-element with the raw data $\vec{r}$. The variable $\vec{w}^*[l]$ is the conjugate of the element of $\vec{w}$ at index l. The variable $\vec{r}[m + l]$ is the element of $\vec{r}$ at index $m + l$. It is assumed that the length of $\vec{r}$, K, is greater than L, the length of $\vec{w}$. The output, $\vec{f}[m]$, is one element in the feature vector $\vec{f}$. It can be seen that the index $m + l$ can extend beyond K (the length of $\vec{r}$) if allowed to run from 0 to $K - 1$. Based on this equation, the length of $\vec{f}$ will be $M = K - L + 1$. A decision metric can be calculated from the output of the correlation, $\vec{f}$.

$$\vec{f}[m] = \sum_{l=0}^{L-1} \vec{r}[m + l]\,\vec{w}^*[l] \tag{5.42}$$

This technique can also be implemented as a **Matched Filter**. Reference [8] provides a thorough derivation of the matched filter impulse response, which we do not replicate here. Here we highlight its conclusions, as also summarized in [17]. Matched Filters are designed to maximize the SNR of a received *symbol*. To that end, the Matched Filter coefficients are set to the time-reversed conjugate of the desired *symbol*. Some standards define a preamble as a sequence of symbols, so the term *symbol* must be clearly defined. For this discussion, the entire sequence of known values will be considered one *symbol*.

The convolution operation for this matched filter is shown in Eq. (5.43) where $g_{MF}[l]$ is a time reversed copy of $\vec{w}^*[l]$, as shown in Eq. (5.44). Note the that the time offset of $L - l$ is necessary to keep the product of the convolution dependent only on samples received in the past.

$$\vec{f}[m] = \sum_{l=0}^{L-1} \vec{r}[m+L-l-1]g_{MF}[l] \tag{5.43}$$

$$g_{MF}[l] = \vec{w}^*(L-l-1) \tag{5.44}$$

As is shown in Eq. (5.42), the known sequence detector implements **Coherent Summation**. This means that the phases of the samples being summed are taken into consideration. For this reason, the known sequence detector is sensitive to synchronization offsets. This sensitivity will be addressed in the calculation of the decision metric.

5.5.2 Calculating the Decision Metric with no Phase Offset

If there is no carrier phase or frequency offset, then the decision metric may simply be $\max(\vec{f})$. If the decision metric is set to $\max(\vec{f})$ then the distribution of the output will be Gaussian, similar to that shown in Figure 5.3. If no signal is present, the distribution of the decision metric will be that of noise, zero-mean Gaussian. If the desired signal is present, then the distribution of decision metric will be Gaussian with a nonzero mean. The function $c(\vec{f})$ is defined in Eq. (5.45). Remember that $c(\vec{f}) = d(\vec{r})$. The maximum of $\vec{f}$ is an amplitude A to which a noise sample n is added. The noise sample n is complex valued, with a real component $\text{Re}(n)$, and an imaginary component $\text{Im}(n)$. The real and imaginary components are independent. Both components follow a Gaussian distribution with zero mean and variance $K\sigma_n^2$. This means that the complex-valued noise sample n_{sum} follows a circularly-symmetric complex normal distribution. In this case, where there is no phase offset, the summation is completely coherent and the amplitude output A is a positive real value. Therefore, only the real component of the noise sample needs to be considered for the decision metric.

$$c(\vec{f}) = \max(\text{Re}(\vec{f})) = A + \text{Re}(n_{sum}) \tag{5.45}$$

From this definition of $c(\vec{f})$ we can now calculate the probability of missed detect in Eq. (5.46), and the probability of false alarm in Eq. (5.47); the threshold is η. Both of these expressions are based on the complementary cumulative distribution of a Gaussian distribution represented by the Q function which is defined in Eq. (5.35).

$$P_{MD} = Q\left(\frac{A-\eta}{\sqrt{K}\sigma_n}\right) \tag{5.46}$$

$$P_{FA} = Q\left(\frac{\eta}{\sqrt{K}\sigma_n}\right) \tag{5.47}$$

Solving for the threshold value given a probability of False Alarm or Missed Detect involves the inverse Q function, Q^{-1}, as shown in Eqs. (5.48) and (5.49). The inverse Q function is available in tools such as *MATLAB*.

$$\eta_{\text{FA}} = \sqrt{K}\sigma_n Q^{-1}(P_{\text{FA}}) \tag{5.48}$$

$$\eta_{\text{MD}} = A - \sqrt{K}\sigma_n Q^{-1}(P_{\text{MD}}) \tag{5.49}$$

5.5.3 Calculating the Decision Metric with a Constant Phase Offset

If there is a constant carrier phase offset, then the samples in $\vec{r}$ will be offset in phase by $e^{i\phi}$. This is shown in Eq. (5.50) where $\vec{r}_\phi[k]$ is a phase-offset version of $\vec{r}[k]$.

$$\vec{r}_\phi[k] = \vec{r}[k]e^{i\phi} \tag{5.50}$$

The features of this raw data offset in phase will be disambiguated from the feature in Eq. (5.42) by using the subscript ϕ. The extraction of the updated features is expressed in Eq. (5.51) by combining Eqs. (5.50) and (5.42).

$$\vec{f}_\phi[m] = \sum_{l=0}^{L-1} \vec{r}_\phi[m+l]\vec{w}^*[l] \tag{5.51}$$

The expression in Eq. (5.51) can be simplified by recognizing that $e^{i\phi}$ does not depend on the index l yielding Eq. (5.52) where the phase offset is placed outside of the summation. The result expresses the linear relationship between $\vec{f}_\phi[m]$ and $\vec{f}[m]$.

$$\vec{f}_\phi[m] = \sum_{l=0}^{L-1} r[m+l]e^{i\phi}\vec{w}^*[l]$$

$$= e^{i\phi}\sum_{l=0}^{L-1} \vec{r}[m+l]\vec{w}^*[l]$$

$$= e^{i\phi}\vec{f}[m] \tag{5.52}$$

The maximum value of $\text{Re}\left(\vec{f}[m]\right)$ is located at the index $\hat{m}$ calculated in Eq. (5.53). The value of $\vec{f}[\hat{m}]$ is given in Eq. (5.54). This is a complex value because the noise sample n is complex. The amplitude value A is real. The index $\hat{m}$ is then used to find the corresponding value of $\vec{f}_\phi[m]$, which is the value of $\vec{f}[\hat{m}]$ phase shifted by $e^{i\phi}$, as shown in Eq. (5.55).

$$\text{argmax}\left(\text{Re}\left(\vec{f}[m]\right)\right) = \hat{m} \tag{5.53}$$

$$\vec{f}[\hat{m}] = A + n_{\text{sum}} \tag{5.54}$$

$$\vec{f}_\phi[\hat{m}] = e^{i\phi}\vec{f}[\hat{m}] = Ae^{i\phi} + n_{\text{sum}}e^{i\phi} \tag{5.55}$$

This now presents a problem. The value $Ae^{i\phi}$ is complex. This means that it is insufficient to take the real component in order to locate the maximum of $\vec{f}_\phi[\hat{m}]$, as was done for $\vec{f}[\hat{m}]$ in Eq. (5.45). Instead, the maximum squared magnitude is found in Eq. (5.56).

$$c_\phi\left(\vec{f}_\phi\right) = \max\left(\left|\vec{f}_\phi\right|^2\right) \tag{5.56}$$

Taking the square magnitude changes the distribution of the decision metric. This in turn changes the probability of a missed detect to Eq. (5.57) and changes the probability of false alarm to Eq. (5.58). The function Q_1 is the Marcum Q function for $K = 1$ and is defined in Eq. (5.31).

$$P_{MD} = 1 - Q_1\left(\frac{|A|}{\sqrt{K}\sigma_n}, \frac{\sqrt{\eta}}{\sqrt{K}\sigma_n}\right) \tag{5.57}$$

$$P_{FA} = e^{\frac{-\eta}{2K\sigma_n^2}} \tag{5.58}$$

A threshold for a given probability of false alarm can be determined using Eq. (5.58), and that is shown in Eq. (5.59). Calculating a threshold for a given probability of missed detect is a bit more difficult as that involves the inverse Marcum Q function.

$$\eta_{FA} = -2\sqrt{K}\sigma_n^2 \ln\left(P_{FA}\right) \tag{5.59}$$

5.5.4 Calculating the Decision Metric with a Constant Frequency Offset

A constant frequency offset is a case where the phase offset has a constant derivative with respect to time. Therefore, the phase offset changes as a function of time and the phase offset applied to samples of $\vec{r}$ may not be the same. This is expressed in Eq. (5.60) where ω represents the constant frequency offset.

$$\vec{r}_\omega[k] = \vec{r}[k]e^{i\omega k} \tag{5.60}$$

The effect of this constant frequency offset on the calculation of the features is shown in Eq. (5.61) by combining Eqs. (5.60) and (5.42). Because the phase offset is a function of time, the term cannot be excluded from the summation.

$$\vec{f}_\omega[m] = \sum_{l=0}^{L-1} \vec{r}[m+l]e^{i\omega(m+l)}\vec{w}^*[l] \tag{5.61}$$

The effect is that the correlation peak becomes a function of this frequency offset. The degradation to the correlation peak is illustrated in Figure 5.6. The y-axis represents decibels of loss to the power of the correlation peak. The x-axis represents the frequency offset relative to the period of correlation, which is the

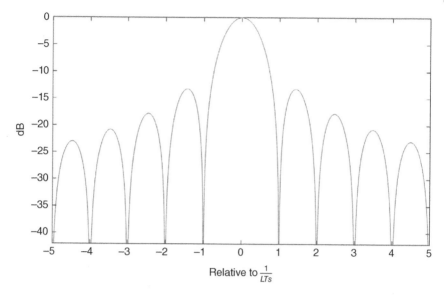

Figure 5.6 Loss due to center frequency offset.

correlation L times the sample period. As is evident, the known sequence detector performance is degraded in the presence of frequency offsets.

5.6 Cyclic Spectral Analysis

5.6.1 Motivation

Energy detection is highly susceptible to noise. Specifically, when the uncertainty of the noise floor estimate exceeds the power of the active signal, that signal becomes indistinguishable from the environment with respect to power spectral density. For similar reasons, energy detection is also highly susceptible to cochannel interference. If multiple signals occupy the same bandwidth simultaneously, they become indistinguishable from each other with respect to power spectral density. This is illustrated in Figure 5.7 where we see two baseband signals, one BPSK and one 16-QAM, separated by only 0.1 Hz (normalized). Given their respective occupied bandwidths, there is significant spectral overlap. If they occur simultaneously, i.e., *collide* for any reason, they become indistinguishable in the power spectrum, as shown in subplot (c).

From a wireless coexistence perspective, we then risk losing important information regarding either signal, which would otherwise contribute to our ability to identify spectrum opportunities similar to those illustrated in Figure 5.1.

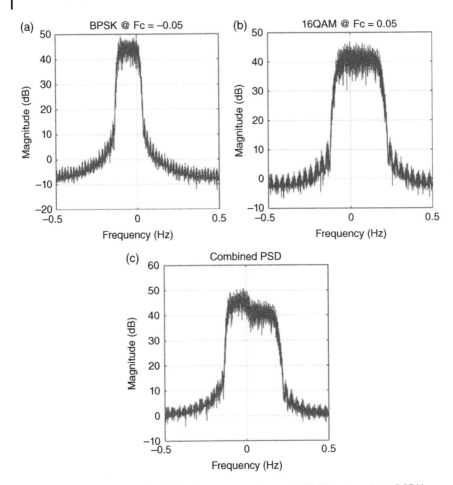

Figure 5.7 Overlapping signals in the power spectrum. (a) BPSK centered at −0.05 Hz (normalized). (b) 16-QAM centered at 0.05 Hz (normalized). (c) Composite spectrum (normalized).

Assuming persistent signals for the moment, energy detection models signal activity as stationary (wide sense), under the assumption that each signal's statistical properties do not vary with time. However, we can instead model signal activity as cyclostationary, under the assumption that each signal's statistical properties are periodic in time. Example periodicities for digital communications signals include carrier frequency and symbol or chip rates. As such, we can apply nonlinear time-invariant transformations, which produce spectral lines indicative of these periodicities when present. This is the basis of cyclic spectral analysis, and is explored extensively in [18].

As these periodicities exhibit little correlation with noise, cyclic spectrum analysis is much less susceptible to noise than energy detection. Furthermore, assuming each periodicity can be mapped to a unique point in the two-dimensional bifrequency plane (spectral vs. cyclic frequencies); cyclic spectrum analysis is much less susceptible to cochannel interference than energy detection.

5.6.2 Spectral Correlation Density

We now develop the function central to cyclic spectrum analysis, the Spectral Correlation Density function (SCD) as defined in Refs. [18–20]. For consistency, we follow the notation used in Ref. [18] as it contains an especially rigorous development of the topics highlighted here. We start by introducing the discrete time Cyclic Autocorrelation Function (CAF) shown in Eq. (5.62); $r[k]$ represents $r(t)$ sampled at k multiples of the sampling period T_s.

$$R_r^\alpha[\tau] = \left\langle r[k] r^*[k-\tau] e^{-i2\pi\alpha k} \right\rangle e^{i\pi\alpha\tau} \tag{5.62}$$

Here the desired nonlinear transformation is simply a linear combination of delay products used to produce spectral lines representative of any second order periodicities present. We refer to these as second order periodicities, because Eq. (5.62) represents a time-invariant quadratic transformation of $r[k]$. $R_r^\alpha[\tau]$ can be thought of as the Fourier coefficient of an additive sinusoidal component of frequency α contained in the integer delay product dictated by τ. For $\alpha = 0$, $R_r^\alpha[\tau]$ reduces to the conventional autocorrelation function shown in Eq. (5.63), hence the term **cyclic autocorrelation function** for $\alpha \neq 0$.

$$R_r^0 = \left\langle r[k] r^*[k-\tau] \right\rangle \tag{5.63}$$

For further insight, $R_r^\alpha[\tau]$ can be expressed in terms of frequency translates $u[k]$, $v[k]$ as discussed in [18]. Recall that multiplying a signal by $e^{\pm i\pi\alpha k}$ shifts its frequency content by $\pm \frac{\alpha}{2}$; hence the term **frequency translates**. Therefore, Eq. (5.62) can be re-written, as shown in Eqs. (5.64)–(5.66) through factoring of $e^{\pm i2\pi\alpha k}$. The importance of this representation will become apparent shortly.

$$R_r^\alpha[\tau] = \left\langle u[k] v^*[k-\tau] \right\rangle \tag{5.64}$$

$$u[k] = r[k] e^{-i\pi\alpha k} \tag{5.65}$$

$$v[k] = r[k] e^{i\pi\alpha k} \tag{5.66}$$

Now, given $R_r^\alpha[\tau]$, we can localize correlation in the frequency domain. To do so, we simply take its Fourier transform to produce the desired SCD function shown in Eq. (5.67).

$$S_r^\alpha[f] = \sum_{\tau=-\infty}^{\infty} R_r^\alpha[\tau] e^{-i2\pi f \tau} \tag{5.67}$$

Per Eq. (5.64), we can interpret Eq. (5.67) as the spectral density of correlation in $u[k]$ and $v[k]$ at frequency f. Recall $S_u[f] = S_r\left[f + \frac{\alpha}{2}\right]$ and $S_v[f] = S_r\left[f - \frac{\alpha}{2}\right]$ per Eqs. (5.65) and (5.66). We can also interpret Eq. (5.67) as the spectral density of correlation in $r[k]$ at frequencies $f \pm \frac{\alpha}{2}$ as Eqs. (5.62) and (5.64) are equivalent. In either interpretation, α clearly dictates the frequency separation between points of correlation for each (f, α) point estimate.

Since $R_r^{\alpha}[\tau]$ is periodic in α with period 2, so too is $S_r^{\alpha}[f]$. Also, since τ is an integer, $S_r^{\alpha}[f]$ is periodic in f with period 1. The domain of α is $[-1, 1]$, and the domain of f is $\left[-\frac{1}{2}, \frac{1}{2}\right]$.

Similar to our analysis of $R_r^{\alpha}[\tau]$, for $\alpha = 0$, $S_r^{\alpha}[f]$ reduces to the conventional power spectral density (PSD) function shown in Eq. (5.68).

$$S_r^0[f] = \sum_{\tau = -\infty}^{\infty} R_r^0[\tau] e^{-i2\pi f \tau} \tag{5.68}$$

For purposes of selecting a detection threshold, we would like $\left|S_r^{\alpha}[f]\right|$ to be bounded to the interval $[0, 1]$. Assuming $u[k]$ and $v[k]$ do not contain spectral lines in their respective PSDs at $f \pm \frac{\alpha}{2}$, we can normalize $S_r^{\alpha}[f]$ by the geometric mean of their respective variances. This is shown in the denominator of Eq. (5.69), where $\rho_r^{\alpha}[f]$ is the normalized SCD function commonly used to detect signal activity. That is to say, given a block of samples r, we first compute $\left|\rho_r^{\alpha}[f]\right|$ for all (f, α) pairs of interest. Then, any point exceeding the chosen threshold can be considered detection of some form of signal activity.

$$\rho_r^{\alpha}[f] = \frac{S_r^{\alpha}[f]}{\left[S_r^0\left[f + \frac{\alpha}{2}\right] S_r^0\left[f - \frac{\alpha}{2}\right]\right]^{\frac{1}{2}}} \tag{5.69}$$

Threshold selection for $\rho_r^{\alpha}[f]$ is not straightforward; in fact, it is common practice to employ empirical approaches to optimize over the desired values of P_D and P_{FA} for a given application [21]. The main reason being that for blind signal detection, we do not have an expression for $\rho_r^{\alpha}[f]$ derived from a common signal model. Furthermore, there are numerous over-the-air impairments which also affect the surface of $\rho_r^{\alpha}[f]$. We can however refer the reader to reference [22], which proposes a closed form solution to threshold setting based solely upon the desired P_{FA} and the number of observations available.

It is worth noting the implicit dependence Eq. (5.69) has on SNR. If the input SNR is too high, the denominator can become unstable due to $S_r^0\left[f \pm \frac{\alpha}{2}\right]$ approaching zero. This may increase P_{FA} due to products in $\rho_r^{\alpha}[f]$, other than second order periodicities, exceeding the chosen threshold.

It is also worth noting that given no *a priori* information, the entire domain of $\rho_r^\alpha[f]$ must be computed. This requires a considerable number of calculations, which may be prohibitive for certain applications and computing platforms. As such, we explore an efficient approximation of $S_r^\alpha[f]$ called the FFT Accumulation Method (FAM) later in this chapter.

5.6.3 Bifrequency Plane

In the previous section, we developed the normalized SCD function $\rho_r^\alpha[f]$ which indicates the presence of any second order periodicities in the domains of f and α. The region of support is a two dimensional surface on the f, α axes commonly referred to as the bifrequency plane. Assuming a bandpass signal with minimum frequency content b and maximum frequency content B, we can express the non-zero regions of $\rho_r^\alpha[f]$ using Eqs. (5.70) and (5.71).

$$\left| |f| - \frac{|\alpha|}{2} \right| \leq b \tag{5.70}$$

$$\left| |f| + \frac{|\alpha|}{2} \right| \geq B \tag{5.71}$$

As such, its support region resembles a diamond with vertices $(f, \alpha) = (0, \pm 1)$ and $\left(\pm \frac{1}{2}, 0 \right)$. This is illustrated in Figure 5.8 for both bandpass and lowpass signals. Subplot (a) illustrates the support region for a bandpass signal; subplot (b) illustrates the support region for a lowpass signal, and is derived by letting b approach zero in Eq. (5.70).

5.6.4 Implementation

5.6.4.1 Cyclic Cross Periodogram

$S_r^\alpha[f]$ for each (f, α) channel pair is commonly estimated via the time-smoothed Cyclic Cross Peridogram (CCP) described in Eq. (5.72). Note the notation change to $S_{rr_T}^{\alpha_0}[k, f_0]$ to reflect that each estimate is with respect to itself, S_{rr} versus S_{xy}, using the specific cyclic and spectral frequencies α_0 and f_0. With respect to implementation concerns, the CCP, as shown is not necessarily more efficient than explicit implementation of Eqs. (5.64)–(5.67). It does, however, provide the necessary framework for the efficiency improvements provided by the FAM method discussed later in this chapter.

Since for any implementation we are forced to work with finite quantities, it makes sense to first define a few variables key to both the CCP and FAM. We again follow the notation found in Ref. [18] for consistency:

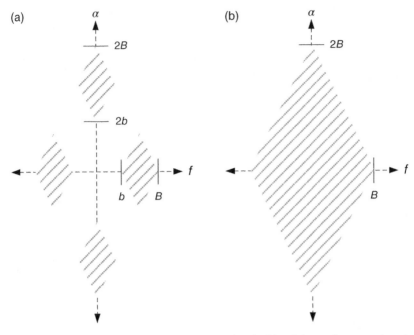

Figure 5.8 SCD support region. (a) Bandpass signal with minimum frequency b and maximum frequency B. (b) Lowpass signal with maximum frequency B.

- Δt represents our entire observation period in seconds, and is composed of N consecutive samples of $r[k]$
- T represents our data tapering period, in seconds, within Δt, and is composed of N' consecutive samples of $r[k]$
- Δf represents the CCP spectral frequency resolution in Hertz, and is calculated as $\frac{1}{T}$
- Δa represents the CCP cyclic frequency resolution in Hertz, and is calculated as $\frac{1}{\Delta t}$

Figure 5.9 illustrates an architecture commonly used to implement the CCP. $r[k]$ is first filtered using one-sided bandpass filters of length N'. Recall that α is the difference between f_1 and f_2; as such, the bandpass filters are centered at $f_0 \pm \frac{\alpha_0}{2}$. The filter shape is dictated by data tapering window $a[n]$, and has a bandwidth Δa on the order Δf. The filter outputs are then decimated by a factor of L, which we assume to be 1 for purposes of this discussion. The decimated sequences are then mixed down to baseband, which then represent the spectral components of $r[k]$, in a bandwidth Δa, centered at $f_0 \pm \frac{\alpha_0}{2}$. These are commonly referred to as complex

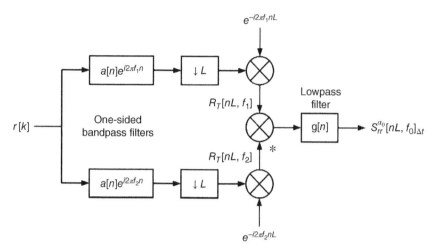

Figure 5.9 Cyclic cross periodogram.

demodulates, and are expressed mathematically by Eq. (5.73). Note that although mathematically equivalent, the order of operations in deriving $R_T[k, f]$ do not match those shown in Figure 5.9. This is because the location of the frequency mixers in Figure 5.9 provides the framework necessary to develop the FAM method described in Section 5.6.4.2.

The complex demodulates $R_T[k, f]$ are then correlated over the time period Δt. This process is implemented as their conjugate product, followed by a lowpass filter. The filter shape is dictated by data tapering window $g[n]$, which has length N and a bandwidth on the order of $\frac{1}{\Delta t}$.

It is worth reinforcing that the CCP architecture shown in Figure 5.9 produces an *estimate* of $S_{rr}^{\alpha_0}[f_0]$. For any real implementation, Δf does not approach zero, and Δt does not approach infinity. As such, $S_{rr_T}^{\alpha_0}[k, f_0]_{\Delta t}$ approximates $S_{rr}^{\alpha_0}[f_0]$, and its accuracy depends heavily upon the design of window functions $a[n]$ and $g[n]$, and the length of the observation period Δt.

$$S_{rr_T}^{\alpha_0}[k, f_0]_{\Delta t}$$
$$= \sum_n R_T[n, f_1] R_T^*[n, f_2] g[k - n] \tag{5.72}$$

$$R_T[k, f] = \sum_{n=-\frac{N'}{2}}^{\frac{N'}{2}-1} a[n] r[k - n] e^{-i2\pi f(k-n)T_s} \tag{5.73}$$

Before concluding our introduction to the CCP, it must be mentioned that for a reliable estimate of $S_{rr}^{\alpha_0}[f_0]$, the time/frequency resolution product $\Delta t \Delta f$ must be much greater than unity. This is equivalent to $\frac{N}{N'} \gg 1$, where the number of samples in the observation interval is much greater than that of the data tapering window. This results in the two-dimensional rectangular point estimate described by Eqs. (5.74) and (5.75), and shown in Figure 5.10. In reality, each point estimate has both a main lobe and side lobes, with the aforementioned rectangle representing only the main lobe. Features falling outside the main lobe, but still within the side lobes, are considered *cycle leakage*. Cycle leakage can still contribute to $S_{rr_T}^{\alpha_0}[k, f_0]$, and is minimized by reducing the side lobes of data tapering window $a[n]$.

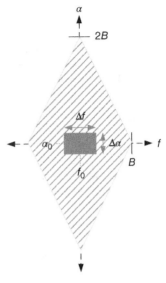

Figure 5.10 CCSP point estimate support region

$$|f - f_0| \leq \frac{\Delta f}{2} \tag{5.74}$$

$$|\alpha - \alpha_0| \leq \frac{1}{2\Delta t} \tag{5.75}$$

5.6.4.2 FFT Accumulation Method

For a limited number of point estimates, the architecture shown in Figure 5.9 has a computational complexity similar to other spectral analysis methods [18]. In other words, if the expected signal activity lies within a known and limited range of spectral or cyclic frequencies, this architecture may be sufficient. In contrast, if the entire bifrequency plane must be estimated, the architecture shown in Figure 5.9 can be quite expensive to run as is. It is not the computation of the spectral components themselves that is costly, but rather the large number of correlation computations required [19]. As such, we would like a more efficient way to estimate the entire bifrequency plane. One such method is the FFT Accumulation Method (FAM), which as the name implies, utilizes the efficiency of the Fast Fourier Transform (FFT) in computing several component functions of the CCP. We develop the FAM algorithm described in Refs. [18, 19] through several modifications to Eq. (5.72).

We first make use of the decimation factor L shown in Figure 5.9. Since the filter outputs are oversampled by N', we can reduce the output sampling rate f_s by a factor of L. We do so by consuming $r[k]$ in groups of L-spaced sample blocks of length N'. Thus, $P = \frac{N}{L}$ samples are processed per point estimate rather than N. Common practice is to set $L = \frac{N'}{4}$ as a good compromise between improving efficiency but still preventing aliasing.

Second, we can compute Eq. (5.73) using an N'-point FFT. This discretizes the translate center-frequencies to $f_m = \frac{m f_s}{N'}$ for $m = \frac{-N'}{2} : \frac{N'}{2} - 1$.

Lastly, we add a discretized frequency shift, $\epsilon = q\Delta\alpha = \frac{q}{P}$ for $|q| < \frac{\Delta a}{\Delta \alpha}$ to the frequency translate conjugate product. This allows us to compute point estimates using a P-point FFT, instead of the explicit time smoothing alluded to in Eq. (5.72). Each FFT generates a vector of P point estimates centered at $f_0 = \frac{f_l + f_m}{2}$ and $\alpha_0 = f_l - f_m + \epsilon$. Regarding its support region, each estimate now has a variable main lobe width of $\Delta f = \Delta a - |\epsilon|$, and a constant height of $\Delta \alpha = \frac{1}{P}$.

Combining all modifications yields a point estimate expressed as Eq. (5.76) and illustrated in Figure 5.11. In all, this is a much more efficient CCP implementation than Eq. (5.72), which is especially beneficial when computing point estimates over the entire bifrequency plane [18].

$$S_{rr_T}^{\alpha_0}[pL, f_{lm}]_{\Delta t}$$

$$= \sum_p R_T[pL, f_l] R_T^*[pL, f_m] g[p - n] e^{\frac{-i2\pi nq}{P}} \tag{5.76}$$

The FAM implementation does however have one side effect deserving attention. Each point estimate is now indexed by $(f_0, \alpha_0 + \epsilon)$, where ϵ represents a discretized frequency shift. As such, the frequency resolution Δf is degraded by ϵ, and approaches zero s ϵ increases. The result is unreliable estimates ($\Delta t \Delta f$ not $\gg 1$) near the edges of the P-point FFT. As such, these are commonly discarded. This artifact certainly does not negate the significant computational cost savings of the FAM implementation, but serves as another reminder of the *no free lunch theorem*.

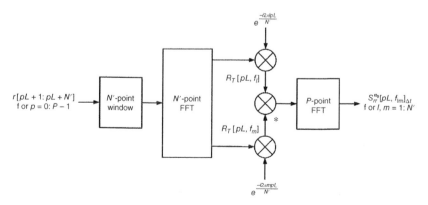

Figure 5.11 FAM implementation of CCP.

5.6.4.3 FAM Design Example

We end our treatment of cyclic spectrum analysis with a design example using the FAM implementation developed in Section 5.6.4.2. We assume no *a priori* information, so our design parameters are simply Δf and $\Delta \alpha$. For this example we choose $\Delta f = 0.01$ (normalized) and $\Delta \alpha = 0.0001$ (normalized). Given these values we derive N', L, P, and N as the following.

$$N' = \frac{1}{\Delta f} = \frac{1}{0.01} = 100 \rightarrow 128 \text{ (power of 2)}$$

$$L = \frac{N'}{4} = \frac{128}{4} = 32$$

$$P = \frac{1}{\Delta \alpha L} = \frac{1}{(0.0001)(32)} = 312.5 \rightarrow 512 \text{ (power of 2)}$$

$$N = PL = (512)(32) = 16384$$

We also see that the time/frequency resolution product $\Delta t \Delta f = \frac{N}{N'} = \frac{16384}{128} = 128 \gg 1$, so we expect our estimates to be reliable. Lastly, we've implemented $a(n)$ as a Blackman window, and $g(n)$ as a rectangular window. The Blackman window, shown in Figure 5.12, provides a good compromise between main lobe width and side lobe suppression in the frequency domain.

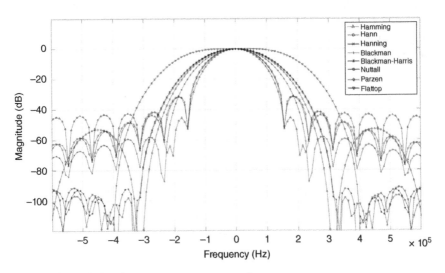

Figure 5.12 Window frequency response comparison.

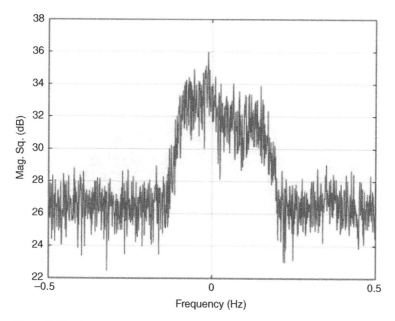

Figure 5.13 Power spectrum of overlapping signals.

Given these design parameters, we revisit the example illustrated in Figure 5.7, two signals which overlap in the spectrum. The first is a BPSK signal centered at $f_c = -0.05$ (normalized), with a symbol rate of 0.125 (normalized); the second is a 16-QAM signal centered at $f_c = 0.05$ (normalized), with a symbol rate of 0.25 (normalized). Both signals have an SNR of 3dB. As is shown in Figure 5.13, these signals overlap heavily with respect to spectral frequency. However, they are separable in cyclic frequency due to their unique symbol rates. This is shown in Figure 5.14, which represents the bifrequency plane for $\alpha > 0$.

5.7 Final Thoughts

The goal of this chapter was to provide theoretical background and implementation details such that the reader could understand the spectrum sensing techniques identified in existing wireless standards. To that end, three signal detectors were discussed in detail. Those being:

- Energy Detector
- Known Pattern Detector
- Cyclic Spectral Analysis

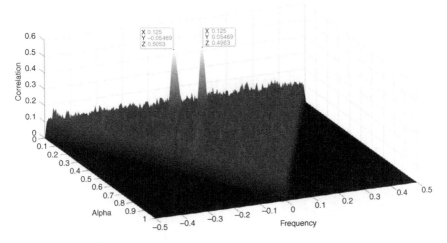

Figure 5.14 Bifrequency plane of overlapping signals.

For the IEEE 802.11 standard, only the matched filter and the energy detector are utilized. The IEEE 802.22 standard calls out many different types of detectors, some falling into these three categories. The list of IEEE 802.22 detectors goes beyond these examples. The different algorithms employed in spectrum sensing in the IEEE 802.22 standard are explored in depth in references [23, 24].

Spectrum sensing for opportunistic spectrum access has much in common with Carrier Sense Multiple Access (CSMA), which was discussed in Chapter 4. Both make heavy use of sensing algorithms. CSMA focuses on the use of one frequency channel, and is a method employed by primary users. The Hidden Node problem described in the CSMA section in Chapter 4 affects both CSMA and spectrum sensing for opportunistic spectrum access. One solution is to spatially distribute the spectrum sensing across multiple nodes. This requires collaboration between the different nodes. Such a solution is employed by the IEEE 802.22 standard and will be discussed in Chapter 8.

References

1 Federal Communication Commission.(2002). Spectrum Policy Task Force. *Report ET Docket 02-135.*

2 Part 22. (2011). Cognitive Wireless RAN Medium Access Control (MAC) and Physical Layer (PHY) Specifications: Policies and Procedures for Operation in the TV Bands, IEEE 802.22-2011.

3 Part 11. (2016). Wireless LAN Medium Access Control (MAC) and Physical Layer (PHY) Specifications, IEEE 802.11-2016, IEEE, 2016.

4 IEEE. (2019). IEEE Standard for Definitions and Concepts for Dynamic Spectrum Access: Terminology Relating to Emerging Wireless Networks, System Functionality, and Spectrum Management, IEEE Std 1900.1-2019, IEEE, 2019.

5 A. Goldsmith, S. A. Jafar, I. Maric and S. Srinivasa, "Breaking Spectrum Gridlock With Cognitive Radios: An Information Theoretic Perspective," *Proceedings of the IEEE*, vol. 97, no. 5, pp. 894-914, 2009.

6 IEEE. (1900). Standards Committee, IEEE 1900.4a, IEEE, 2001.

7 IEEE Std 6-2011. (1900). IEEE Standard for Spectrum Sensing Interfaces and Data Structures for Dynamic Spectrum Access and other Advanced Radio Communication Systems, IEEE, 2011.

8 Sklar, B. *Digital Communications: Fundamentals and Applications.* Englewood Cliffs: Prentice Hall.

9 McDonough, R.N. and Whalen, A.D. (1995). *Detection of Signals in Noise.* San Diego: Academic Press, Inc.

10 Arjoune, Y., El Mrabet, Z., El Ghazi, H. et al. (2018). Spectrum Sensing: Enhanced Energy Detection Technique Based on Noise Measurement. *IEEE 8th Annual Computing and Communication Workshop and Conference*, Las Vegas, 2018.

11 R. R. Nadakuditi and A. Edelman, Sample eigenvalue based detection of high-dimensional signals in white noise using relatively few samples. *IEEE Transactions on Signal Processing* 56, no. 7, 2008.

12 Sarker, M. (2015). Energy detector based spectrum sensing by adaptive threshold for low SNR in CR networks. *24th Wireless and Optical Communication Conference (WOCC)*, Taipei, 2015.

13 Russell, S. and Norvig, P. (2010). *Artificial Intelligence: A Modern Approach.* Prentice Hall: Upper Saddle River.

14 H. Urkowitz, Energy detection of unknown deterministic signals *Proceedings of the IEEE*, vol. 55, no. 4, 1967.

15 F. Digham, M. Alouin and M. Simon, On the energy detection of unknown signals over fading channels *IEEE Transactions on Communications*, vol. 55, no. 1, pp. 21-24, 2007.

16 Cabric, D., Tkachenko, A., and Brodersen, R.W. (2006). Experimental study of spectrum sensing based on energy detection and network cooperation. *Proceedings of the First International Workshop on Technology and Policy for Accessing Spectrum*, 2006.

17 Lyons, R.G. (2010). *Understanding Digital Signal Processing.* Upper Saddle River, NJ: Prentice Hall.

18 Gardner, W. (1994). *Cyclostationarity.* New York: IEEE Press.

19 W. A. Brown and H. H. Loomis, Digital implementations of spectral correlation analyzers *IEEE Signal Processing Magazine*, vol. 41, no. 2, pp. 703-720, 1993.

20 R. S. Roberts, W. A. Brown and H. H. Loomis, Computationaly efficient algorithns for cyclic spectral analysis *IEEE Signal Processing Magazine,* pp. 38-49, 1991.

21 Spooner, C. (2016). The spectral correlation function, 19 June 2016. https://cyclostationary.blog/2015/09/28/the-spectral-correlation-function/ (accessed 23 December 2020).

22 C. Carter, "Receiver operating characteristics for a linearly thresholded coherence estimation detector," *IEEE Transactions on Acoustics, Speech, and Signal Processing,* vol. 25, no. 1, pp. 90-92, 1977.

23 H. Kim and K. G. Shin, "In-band spectrum sensing in IEEE 802.22 WRANs for incumbent protection," *IEEE Transactions on Mobile Computing,* vol. 9, no. 12, pp. 1766-1779, 2010.

24 Shellhammer S.J. (2008). Spectrum Sensing in IEEE 802.22. *IAPR Workshop on Cognitive Information Processing*, Santorini, Greece, 2008.

6

Intelligent Radio Concepts

6.1 Introduction

6.1.1 Motivation

In this chapter, we introduce intelligent radio concepts. *What is an intelligent radio? How does an intelligent radio improve wireless coexistence? What are its major components and functions?* We answer these questions in part by reviewing a considerable body of research. This includes the concepts and ontology developed in the IEEE Dynamic Spectrum Access Networks Standards Committee (DySPAN-SC) 1900 series [1–3], which is widely considered the first set of standards published specifically targeting cognitive radio and cognitive wireless networks. We also formulate our own forward-looking use-cases describing how intelligent radio concepts can be used to mitigate the wireless coexistence challenges discussed in selected references and chapters of this text.

6.1.2 Definitions

We first establish a set of basic definitions describing the various radio technologies associated with DSA. These and others are treated extensively in [1–3]; we summarize them here, and illustrate their relationships in Figure 6.1.

- **Software-defined radio** – a type of radio in which some or all of its physical layer (PHY) functionality is software-defined, i.e. can be changed without physical modification after manufacture.
- **Software-controlled radio** – a type of radio where some or all of its interface functions and parameters can be set or managed by software.
- **Policy-based radio** – a type of radio in which its behavior is governed by a policy-based control mechanism.

Wireless Coexistence: Standards, Challenges, and Intelligent Solutions, First Edition.
Daniel Chew, Andrew L. Adams, and Jason Uher.
© 2021 The Institute of Electrical and Electronics Engineers, Inc.
Published 2021 by John Wiley & Sons, Inc.

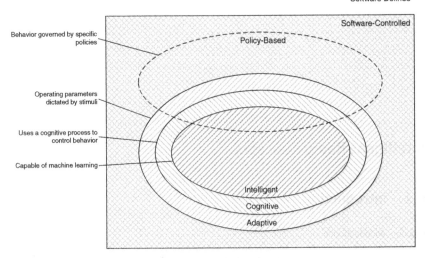

Figure 6.1 Relationship between radio technology terms.

- **Adaptive radio** – a type of radio that has a means of varying its operational parameters in response to internal or external stimuli.
- **Cognitive radio** – a type of radio that utilizes cognitive processes to control its behavior.
- **Intelligent radio** – a type of radio that is capable of machine learning.

From Figure 6.1, we first see that software-controlled radio is a subset of software-defined. This is because the notion of software-control includes a level of operational flexibility not implicit in software-defined. For example, any number of functions could be implemented in software, i.e. "software-defined", but could also only be modified through factory reprogramming; hence the distinction. We next notice the adaptive/cognitive/intelligent hierarchy of software-controlled radios. This makes sense as an adaptive radio, not necessarily a software-controlled radio, has the ability to modify its operating parameters in response to internal or external stimuli. This generally assumes a closed-loop process that does not include a human operator. A cognitive radio implements adaptability through some type of cognitive process. We will discuss the cognitive cycle at length in the next section, but this generally involves the process of observe-orient-decide-act (OODA). Finally, an intelligent radio has the additional capability of machine learning, i.e. it is capable of learning new concepts through observation only.

For the remainder of this text, we focus our attention on intelligent-radio rather than cognitive-radio. The reason being that machine learning is no longer considered a niche capability, but a core competency for many modern software architectures. This is due in large part to the tremendous advancements in open-source development tools. Environments such as TensorFlow [4], Caffe [5], PyTorch [6], Microsoft Cognitive Toolkit [7], and Scikit-Learn [8] allow less experienced engineers to approach machine learning top-down using powerful algorithm libraries and visualization tools.

6.2 Intelligent Radio Use-Cases

In his seminal work defining cognitive radio [9], Mitola describes four use-cases where radios with cognitive capabilities could enhance the ability of wireless infrastructure to utilize the spectrum more efficiently than was done at the time. The motivation being that static spectrum allocations by governing bodies hinder the economic benefit of technological advancements; cellular technology being his exemplar. He argues that social pressure then mounts on governing bodies to reallocate under-utilized spectrum to areas currently experiencing over-use. While this may temporarily relieve pressure, it does not address the underlying issues related to dynamic spectrum usage. In contrast, a communications system that is context and environment aware provides a much better solution. Here, machine-to-machine (M2M) exchange between the handset and infrastructure allows each user to negotiate an appropriate amount of bandwidth needed to access to the spectrum.

The first use-case described in [9] is spectrum pooling. This is an arrangement under which owners of allocated spectrum "rent" their allocations to others needing temporary access. The duration can be as short or as long as needed, so long as the renter is able to immediately release the spectrum to legacy users and critical services such as radar, navigation, and public safety.

The second use-case is the exchange of user traffic patterns with cellular infrastructure. This facilitates the ability of the infrastructure to both improve quality of service (QoS) and maximize network access through an understanding of each user's activity pattern provided by the handset.

The third use-case is enhanced delivery of services to a given user. Given minimal user information, such as a travel itinerary or desired product or service, the handset then gains access to the appropriate wireless infrastructure and provides the user any information relevant to their current circumstance. User information could be relayed explicitly, such as a verbal request or typed question, or inferred from the current communications traffic. The handset also prompts the user for any additional information, personal or otherwise, required to access relevant information.

The final use-case described in [9] is type-certified downloads. Given the service upgrade and maintenance capabilities facilitated by software-defined radios, there is great potential for incompatible software components inadvertently being downloaded into a handset. A self-aware handset has knowledge of its hardware and software limitations, and prevents such occurrences by refusing incompatible software components.

The use-cases described by Haykin in [10, 11] espouse the much broader goals of (i) highly reliable communications whenever and wherever needed, and (ii) efficient utilization of the radio spectrum. These were primarily driven by the conclusions reached in the 2002 FCC Spectrum Policy Task Force Report [12], which found that in general, the lack of available spectrum is driven by poor utilization rather than physical scarcity. In addition, this poor utilization includes both access techniques and federal policy. The following statements, taken directly from the executive summary, provide additional context. *"In many bands, spectrum access is a more significant problem than physical scarcity of spectrum, in large part due to legacy command-and-control regulation that limits the ability of potential spectrum users to obtain such access...advances in technology create the potential for systems to use spectrum more intensively and to be much more tolerant of interference than in the past."*

The use-cases described in [13] include DSA and interoperability. Similar to the spectrum pooling method described in [9], Wyglinski et al. envision DSA as a method to improve utilization of licensed spectrum. He cites significant increase in demand for new wireless services, but also acknowledges the deficiencies existing spectrum regulations have in supporting these services. Again, opportunistic use of licensed spectrum, with deference to licensed users, is the suggested solution. Their interoperability use-case refers to the diverse communications protocols deployed across municipalities and federal agencies. As such, first responders are often unable to communicate with federal agencies during emergencies; hurricane Katrina is his exemplar. An environment aware handset would sense the various protocols, and reconfigure as needed to maintain communications.

Marshall [14] expands the desired capability of DSA from interference-free communications to interference-tolerant communications. This trait is highly desirable in unlicensed spectrum as an ideal spectrum hole, i.e. sufficient bandwidth and time vacancy, is not always available. As such, being able to select a communications configuration both tolerant of and unobtrusive to existing traffic would certainly improve spectrum utilization.

DySPAN-SC [2, 15] describes three use-cases to motivate their proposed architecture: dynamic spectrum assignment, dynamic spectrum sharing, and distributed radio-resource usage optimization. As was discussed in Chapter 2, IEEE 1900.4 targets heterogeneous wireless environments comprised of multiple composite wireless networks (CWN) and multiple user-terminals. Each CWN consists of a packet-based core network, and one or more radio access networks (RAN) supporting one or more radio access technologies (RAT). For example, if 4G, WiFi, and WiMAX were operating simultaneously within a given area, IEEE 1900.4 proposes to optimize the aggregate spectrum usage. It attempts to do so through distributed decision-making, enabled by frequent information exchange. By design, the various RANs within the participating networks are assigned frequency bands appropriate to their current traffic and QoS requirements. Furthermore, RANs can operate in overlapping frequency bands if mutual interference is below tolerable levels. Networks and terminals also collaborate to optimize radio-resource usage. The participating networks determine global operating constraints and communicate these to the terminals being serviced. The terminals then optimize their own resource usage against these constraints, and communicate reconfiguration decisions back to the CWN. We review the proposed IEEE 1900.4 architecture later in this chapter.

As part of Third Generation Partnership Project (3GPP) Release 15, 3GPP executed a feasibility study of Long Term Evolution (LTE) License-Assisted Access (LAA) to unlicensed spectrum [16]. Their use-case deals with meeting the increase in both users of wireless services and individual user's bandwidth requirements. To corroborate their concerns, reference [17] provides growth projections for 3GPP fourth generation (4G) and fifth generation (5G) mobile traffic through 2025. The study focuses on the possible solution of offloading licensed traffic on to unlicensed spectrum when possible. However, similar to spectrum pooling and DSA, LTE must facilitate fair-use with legacy users such as WiFi and Bluetooth. Therefore, LTE LAA must be both context and environment aware while operating in unlicensed spectrum.

More recently, Bhandari and Joshi [18] propose to maximize spectrum utilization by fifth-generation (5G) wireless cellular networks through increased network density. As compared to 4G networks, 5G is expected to provide higher data rates, improved connectivity, lower end-to-end latency, and higher system capacity; all while operating on similar licensed spectrum allocations. Their proposed solution is to create a hierarchy of network access, whereby macro-cells are considered the primary users (PU) of licensed spectrum, and small-cells considered secondary users (SU). This heterogeneous network scenario necessarily requires small-cell deployments to employ intelligent radio functionality to access the spectrum without interfering with macro-cell operation. Their ability to do so is what potentially maximizes the spectrum utilization within the various licensed cellular bands.

It should come as no surprise that many of the wireless services envisioned by Mitola have in fact become reality. Multiple smart-home and smart-phone products now include voice-activated digital assistants. Examples include Amazon

Alexa [19], Microsoft Cortana [20], Google Assistant [21], and Apple Siri [22]. These utilities are used for everything from simple appointment reminders to ordering goods and services. Also prevalent today are bandwidth hungry business applications such as Webex [23] and Zoom [24], and streaming entertainment services such as Netflix [25], YouTube [26], and wireless gaming. Applications such as Google Maps [27], Waze [28], and Mapquest [29] provide walking or driving directions from any compatible GPS enabled wireless device. More importantly, they alert the user to heavy traffic or obstructions and provide alternate routes while in use. Lastly, smart-phone software updates to operating systems and applications are now routinely pushed over WiFi, with minimal interruption in service. Per [30], an Apple iPhone user receives on average eight updates per year over a four year period, which historically is the life-cycle of any particular model [31].

Unfortunately, utilization efficiency has not kept pace with spectrum demand. The utilization problems espoused in [12] persist today, but with greater demand for bandwidth to support the aforementioned applications. Furthermore, many argue that technological advancements could be impeded by lack of spectrum access, regardless of any anticipated social benefits. Lastly, interoperability across protocols, service providers, and wireless infrastructures is not only a safety issue as discussed in [13], but also contributes to the perceived scarcity of spectrum. We therefore look to improve spectrum utilization through *intelligent* methods, which apply equally to both licensed and unlicensed spectrum. For licensed spectrum, we look to improve the utilization of spectrum in order to improve quality of service (QoS) and support new users and services. For unlicensed spectrum, we look to improve the utilization of unlicensed bands by improving our ability to coexist with other unlicensed users. We consider both cases to be wireless coexistence challenges as spectrum utilization is improved only through increasing the occupancy rate of a given band of frequencies.

With these goals in mind, we now list the desired capabilities of an intelligent communications system in the context of wireless coexistence.

- **Be environment aware:** Using local and shared observations, build a predictive model of the spectrum describing all actors, service providers, and infrastructure observed. Use this to identify all possible options for intra-network traffic. These include white spaces, negotiated access to existing infrastructure, and inter-network traffic exchange. The greater the model resolution, the more flexibility afforded the ensuing decision making process.
- **Be context aware:** Determine spectrum access requirements based upon the current communications context, historical traffic patterns, and any anticipated information or service needs. This allows the ensuing decision making process to obtain appropriate spectrum access as needed, such that the user does not experience a gap in delivery of services. We must however acknowledge the significant privacy concerns associated with providing a handset and/or service

provider this much personal information. Unintended consequences include something as benign as product solicitation to something as damaging as the theft of intellectual property or identities [32]. As such, cognitive and intelligent radio research should include methods to protect user-data while also using it to improve QoS.

- **Be self-aware:** It is not enough to be environment aware; to be interoperable with the many standards and protocols observed in a given environment, a handset must be capable of modifying its own configuration to match existing traffic and infrastructure. It also must know its own limitations such that it does not choose a configuration beyond its physical limitations. Self-awareness also facilitates self-optimization. Cognitive radio was envisioned to execute more than DSA algorithms; it was also meant to self-optimize its communication protocol against user needs, available spectrum, and channel conditions.

- **Learn wireless coexistence strategies:** We chose to focus on intelligent rather than cognitive radio concepts [1] because we can harness any learning capacity to perform joint optimization against DSA and overall spectrum utilization. Reference [16] recognizes these competing goals, and addresses them with fair-use policies. However, these policies likely differ across spectrum bands, population density, and geographic regions. As such, treating this as a learning task should ultimately provide a better solution unencumbered by static policies.

It should be clear that being environment, context, and self-aware facilitates the learning of a coexistence strategy for a given situation. This is not to say learning tasks cannot be executed with incomplete information; quite the opposite in fact. However, complete awareness significantly improves the chances of finding the *best* solution vs. *a* solution, which is the goal of an optimization task.

6.3 The Cognitive Cycle

As mentioned earlier, [1] defines a cognitive radio as "*a type of radio that utilizes cognitive processes to control its behavior.*" To implement such processes in software, as defined by DySPAN-SC, one must have a basic understanding of reasoning, i.e. how an animal or human transitions from perception to purposeful action [33]. The reasoning method promoted here is **reasoning by abduction** [33–35], as introduced by Charles Sanders Pierce. Prior to his work, abduction, deduction, and induction were widely considered different methods of reasoning. Pierce argued they are simply different stages of scientific inquiry [35], with abduction being the first. This theory is best illustrated by Pierce's cycle of pragmatism shown in Figure 6.2.

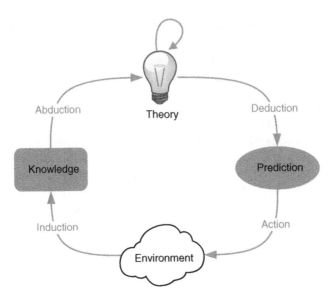

Figure 6.2 Pierce's cycle of pragmatism. *Source:* Modified from Sowa [33].

Looking closer at Figure 6.2, we see the (4) pillars of pragmatism: abduction, deduction, action, and induction. In first person, **abduction** represents our adoption of an explanatory hypothesis, i.e. an inference of a new schema based upon our accumulated knowledge. From abduction, we develop theories, which lead to predictions through **deduction**. We then take *actions*, whose observable effects either prove or disprove our theories through **induction**. This is clearly an iterative process, whereby we continually refine our theories as we interact with our environment. As such, **learning** is the process of accumulating knowledge and formulating *and re-formulating* these into theories [33].

There are many contemporary examples of pragmatism being used for problem solving [36]. For example, the OODA loop developed by John Boyd is a common decision-making model for military command and control (C2) [37]. Originally conceived for air-to-air combat, it closely resembles Pierce's pragmatism model in its use of context and accumulated knowledge in forming hypotheses, which dictate actions. Another example is the Kalman filter used to track moving objects [38, 39]. Given that each track is based upon an object's previous location and current telemetry, this process mirrors pragmatism's ability to refine theories. Lastly, social network analysis is used to identify relationships between specific people and entities [40]. These are typically modeled using graphs, where relationships comprise the edges between nodes. Inferring these relationships through various clustering-by-similarity methods parallels hypothesis generation through abduction [36].

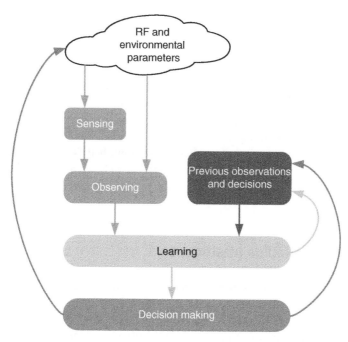

Figure 6.3 Intelligent radio cognitive cycle. *Source:* Abbas et al. [41]. Licensed Under CC BY-4.0.

For application to intelligent radio, we can redraw the pragmatism cycle to reflect our desired capabilities of being environment-aware, self-aware, and learning coexistence strategies. This is shown in Figure 6.3 as suggested in [41]. Abbas et al. does not explicitly link their described learning process to pragmatism; we however see enough similar functionality to suggest this is another contemporary example.

The cognitive cycle starts with observations. For environment-awareness, this also includes some form of sensing to characterize existing activity and identify potential holes for additional traffic. For context-awareness, observations include user and application requirements. For self-awareness, observations include waveform configuration, channel conditions, link status between other network entities, power consumption, and geographic location.

The learning process consists of analyzing current observations, and then forming hypotheses using knowledge of previous observations and decisions. Given multiple learning objectives such as required user throughput, QoS levels, spectrum utilization, and minimum power consumption, multiple hypotheses on how best to achieve each objective are formed. These may in-fact conflict with one another, and as such, optimization must be performed within a given trade-off space.

From hypotheses, decisions are made to take specific actions in support of learning goals. Examples include network maneuvers to mitigate interference, waveform re-configuration to maximize throughput or to be compatible with another network or infrastructure, and resource re-allocation to maximize spectrum utilization. These actions then prompt observations that either support or disprove previous hypotheses, which is the basis of learning.

Clearly, it is of great importance to capture information from which we can form hypotheses and make decisions. Furthermore, we would like to accumulate this information over time, such that it either reaffirms our beliefs, or causes us to infer new schema. How this is accomplished should be weighed against access speed, memory requirements, and information content for various learning algorithms.

6.4 Making Radios Intelligent

Armed with a working knowledge of the cognitive cycle, the question then becomes *"how to implement this in a radio or any other machine?"* This is in fact the focus of the rest of this chapter; but first, let us define a few terms we have all heard many times: automation, machine learning, and machine intelligence. These may seem obvious to some but are often, incorrectly, used interchangeably in much of the literature.

First, automation is neither machine learning nor machine intelligence [42]; it is simply the act of a machine being able to complete a given task without human intervention. There are numerous examples in modern society [43]: smart home appliances, web-based application services, and manufacturing robotics. These all produce the desired outcome, but do not necessarily gather new information for use in future tasks.

Machine learning goes further. It looks for patterns and trends within a given input stream, and uses these to draw conclusions. These conclusions are used solve problems faster and more efficiently than was accomplished before. Examples include voice interactive services, medical imaging devices, and autonomous vehicles.

Machine intelligence is currently at the forefront of technology. By definition, it involves deductive logic, similar to that described within pragmatism. It allows machines to learn from experience; that is, to learn new concepts, realize when it has made a mistake, and correct its behavior to prevent similar mistakes in the future. An intelligent machine requires only high-level programming. It harnesses both automation and machine learning to go beyond executing tasks and to achieve specified goals.

The feasibility of intelligent radio is grounded within these concepts. The use-cases described earlier reiterate our desire to have wireless infrastructure provide many different services regardless of location and circumstance. As such, we expect our devices to be intelligent enough to figure out a way to do so automatically. For both the device manufacture and service provider, this is a significant technical challenge, which has consumed multiple wireless technology related industries for decades.

6.5 Intelligent Radio Architectures

The implementation of cognitive capability is very much an active area of research. There are a litany of challenges to overcome, each being their own vein of research. First, many of the events observed occur on different timelines. For example, changes in the environment such as the channel characteristics between two nodes, or emitter activity in a specific band, may occur on a microsecond scale. In contrast, changes in user applications, network configuration, or geographic location occur on much longer timescales. Furthermore, the observations themselves are not free. For battery powered systems, basic tasks such as spectrum sensing and communicating with other nodes incur significant cost in terms of power consumption. The challenge is to know the value of existing information, such that cost of obtaining new information can be weighed against current goals.

Another implementation challenge is the representation of information. Recall our desire to be self-aware to the extent of being compatible with multiple wireless standards. This necessarily means that any stored information must be agnostic to the configuration in which it was acquired. Otherwise, knowledge becomes specific to each wireless standard; meaning for example, that what was learned while configured for 4G or 5G, cannot be used when configured for 802.22. Pragmatism involves *learning from experience*, regardless of configuration changes.

Where the cognitive capability resides within a particular network is also of concern. Many early implementations chose a **centralized** approach where information is forwarded to a *master* node for learning and decision-making. This master node then processes the given information and informs the rest of the network accordingly. Recent architectures explore a **distributed** approach, where each node learns on its own, and then shares its knowledge with the rest of the network. While the former approach may be easier to implement, it carries significant communications overhead, increased latency in decision-making, and the vulnerability of a single point of failure. The latter, in-theory, increases learning potential, decreases decision-making latency, and is much more resilient to changes in network topology. How to best to realize this however, is an open research question.

With these challenges in mind, we explore the Cognitive Resource Manager (CRM) framework described in [13, 44, 45]. As discussed in [13], we also make no claim to its superiority over other cognitive capability implementations. We do however recognize the similarity between its behavioral model and pragmatism. Beyond that, it is scalable and highly flexible, with significant potential for cross-layer optimization. These combined make this an attractive architectural approach for intelligent radio implementation. We then compare the CRM framework to the functional architecture proposed by IEEE 1900.4 in [2]. In our opinion, these two comprise a sizeable portion of the recent works published on cognitive radio frameworks.

6.5.1 Cognitive Resource Manager Framework

Figure 6.4 illustrates the CRM framework. Originally developed by Matrina Petrova and Petri Mahonen, it is a software component-based architecture comprised of a CRM core, a toolbox of modeling and learning algorithms, and a knowledge database. The core can be considered a micro kernel, which services the information requests made by the various layers of the protocol stack. It does this through modeling of, and learning from the information accumulated in the knowledge database. The core executes modeling and learning tasks using the algorithms available in the toolbox. There is no single way to learn, so the breadth of the toolbox scales with the desired learning capabilities. The knowledge database is continually updated with the information provided to the core; this includes both internal and external stimuli. The policy database is a separate component as shown, but used heavily by the core for decision-making purposes.

What is immediately noticeable is that each layer of the stack has access to the CRM core, and thus the entire knowledge database and toolbox. This facilitates information exchange beyond adjacent layers only, and provides a bevy of learning algorithms to each layer. This does not mean that the OSI model of independent layer functionality is broken; it does however provide access to additional information and tools used for optimization purposes. For example, in the CRM framework, the transport layer has access to link quality metrics collected by the physical layer. This way the TCP does not mistake poor channel conditions for packet collisions, which may cause it to unnecessarily reduce throughput. Furthermore, it can utilize a predictive model of link quality to help determine an effective throughput rate. This ability to share information and learning capabilities across layers is what facilitates improved cross layer optimization.

The CRM framework is capable of supporting both centralized and decentralized network configurations. If centralized, the CRM framework resides on the master node, providing learning and decision making for a collection of

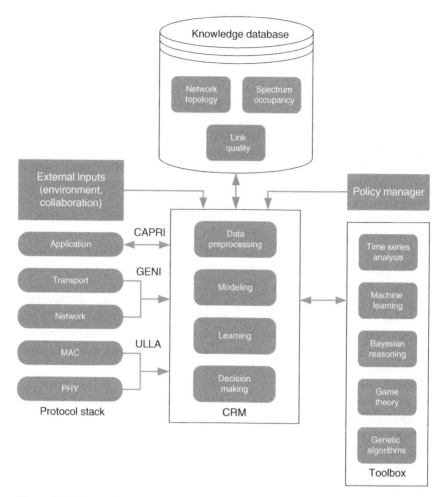

Figure 6.4 Cognitive resource manager framework.

nodes. Here the information being processed is collected from other nodes over some form of communications link. As mentioned above, this carries with it significant communications overhead, as a reliable link is required to maximize available information, and minimize decision-making latency. For de-centralized configurations, the CRM framework resides on each node. Here it processes information collected locally, and is able to share learning outcomes with the entire network. It can also collaborate with adjacent nodes on distributed processing tasks, or determine which node has the lowest cost in performing an individual task.

We now take a closer look at the desired functionality of each component of the framework.

6.5.1.1 CRM Core

The CRM core facilitates the construction and management of the software components, which optimize the various layers of the communications protocol stack. From an implementation point of view, this resembles a micro kernel, providing instantiation, scheduling, and time-synchronization of the various modeling and reasoning functions available within the toolbox. Although only OSI-defined layers are shown in Figure 6.4, we envision platform-level applications also needing access to the core for autonomy purposes.

Given the potentially wide variation in information time-lines, the scheduler needs to support both high-resolution timing and long time-lines, i.e. it must be able to react quickly to new observations, and also prompt the stack for information and actions based on past observations. In this sense, the CRM core also resembles a real-time operating system (OS).

In practice, the core will likely be operating on large quantities of noisy data. Terrestrial environments rarely consist of additive white Gaussian noise (AWGN) only, and many unconditioned time-series exhibit bias and/or outliers. As such, it must employ various data filtering techniques to deal with both linear and non-linear noise. It is in our interest to populate the knowledge database with the best information possible.

As information accumulates in the knowledge database, the core should develop predictive models of various metrics used by the protocol stack. These can then be used as a priori knowledge for future learning and decision-making tasks. For example, given enough observations, the core could instantiate a model predicting spectrum occupancy over the next hour; the model could then be used by the physical layer to minimize interference from other networks.

As alluded to earlier, the information stored in the database must be agnostic to both the wireless standard and the protocol from which it was acquired. As discussed in [13, 44, 45], this can be achieved with unified programming interfaces. These are essentially abstraction layers, which consolidate multiple third party API's into a single data model. Specifically, [45] demonstrates the use of several open interfaces used to abstract LTE functionality from their CRM implementation: Universal Link-Layer API (ULLA) [44, 46] for the PHY and MAC layers, Generic Network Interface (GENI) for the transport and link Layers, and Common Application Requirement Information (CAPRI) [44] for the application layer. This is illustrated in Figure 6.4 as the interfaces between the core and protocol stack.

6.5.1.2 CRM Toolbox

The CRM toolbox is a library of algorithms used by the CRM core for data pre-processing, modeling, learning, and decision-making. From an architectural point of view, this separates the machine learning and optimization modules from the kernel components within the core. The scope of the toolbox reflects the cognitive capability enacted by the core. For example, if the CRM core is expected to generate predictive models from historical data, algorithms which perform time-series analysis are required. If the core is expected to perform multi-dimensional or multi-objective optimization, algorithms which can quickly explore a solution space are required; examples include genetic algorithms and simulated annealing. If the core is expected to reason on incomplete information, algorithms which deal with uncertainty are required; examples include Bayesian reasoning. References [41, 47] provide a good overview of various learning techniques applied to select communications problems. We discuss select algorithms in-depth later in this chapter.

The ability to tailor the toolbox to either the desired network capabilities or platform resource limitations contributes to flexibility and scalability of the CRM framework. For centralized learning and decision-making, the master node may be given significant computing resources, which can support a rich set of toolbox algorithms. For de-centralized operation or low size, weight, and power (SWaP) platforms, the toolbox may be scaled back to those algorithms deemed essential for a given deployment. In either case, the toolbox can grow over time as new numerical methods are discovered.

6.5.1.3 Knowledge Database

The knowledge database provides storage for all the data and models being managed by the CRM core. These include current layer configurations, resource allocations, current observations, historical data, and active models. This represents a substantial amount of information, which must be accessible to the CRM core. As such, from an implementation point of view, access speed and storage capacity should be designed to match expected core capability.

6.5.1.4 Policy Database

The policy database manages the rules and policies set by different stakeholders, and resolves any conflict between them. By stakeholder, we mean any person or body, which has interest in the fair, efficient use of any class of device. These include the user, manufacturer, service provider, or spectrum regulating body. As such, the policies enforced can be user preferences, device limitations, and of course spectrum-use regulations. Similar to the knowledge database, these must be stored in a form agnostic to any wireless standard or protocol layer. It is true that this component is used exclusively by the CRM core, and could functionally be considered part of it; however, for implementation purposes this should be kept separate such that it can be ported or modified without affecting the core.

6.5.2 IEEE 1900.4

6.5.2.1 System Architecture

IEEE 1900.4 attempts to standardize the architecture and interfaces necessary to facilitate optimization of spectrum usage within heterogeneous wireless environments [2, 15]. It does so by specifying the overall system architecture, component interfaces, and information exchange required to perform dynamic channel assignment, dynamic spectrum sharing, and distributed radio-resource usage optimization across supporting CWNs. Working under the assumption each CWN is deployed to service compatible user-terminals within range, optimization of each individual network is less efficient than optimization across all participating networks. As such, information is exchanged between networks and terminals to the extent all are context-aware, and can individually optimize spectrum and radio-resource usage.

The (7) entities that comprise the proposed architecture are illustrated in Figure 6.5. From the network point of view, (4) distinct entities exist: operator spectrum manager (OSM), RAN measurement collector (RMC), RAN reconfiguration controller (RRC), and network resource manager (NRM). The NRM manages the CWN with respect to network-terminal distributed optimization of spectrum usage; the OSM provides it any operator (service provider) specified usage preferences. The RMC provides the NRM all RAN context information, and the RRC executes any RAN configuration changes requested by the NRM. On the terminal side, the TRM manages the terminal with respect to network-terminal distributed optimization, but still within the overall framework dictated by the NRM. Similar to the network side, the TMC provides the TRM terminal context information, and the TRC executes any terminal configuration changes requested by the TRM.

6.5.2.2 Decision Making

Unlike the CRM framework, decision making here is solely policy-based. Two types of policies are defined, spectrum assignment and radio resource usage. Spectrum assignment polices are based upon regulatory requirements, and are generated elsewhere in the network; any operator preferences are captured within the OSM. The NRM uses these policies, and all available RAN context information for dynamic spectrum assignments, determination of spectrum sharing opportunities, RAN selection, and generation of radio-resource usage policies. The NRM also monitors the efficiency of existing spectrum assignments and radio-resource usage

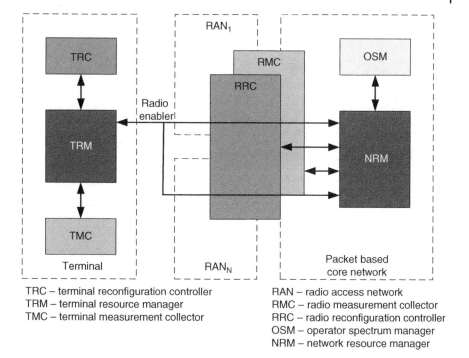

TRC – terminal reconfiguration controller   RAN – radio access network
TRM – terminal resource manager   RMC – radio measurement collector
TMC – terminal measurement collector   RRC – radio reconfiguration controller
     OSM – operator spectrum manager
     NRM – network resource manager

Figure 6.5   IEEE 1900.4 system architecture. *Source:* Buljore et al. [15]. (© [2020] IEEE).

policies as input into future decision-making. The NRM sends configuration changes to the RRC for execution, which in-turn forwards reconfiguration information to the OSM through the NRM. To support distributed decision-making, the NRM sends radio-resource usage policies to the TRM over the NRM-TRM interface. This was envisioned as a logical channel, and is often labeled the Radio Enabler.

On the terminal side, the TRM uses the NRM radio-resource usage policies, and TMC context information for RAN selection and radio-resource usage optimization. The TRM sends configuration changes to the TRC for execution, and returns reconfiguration information back to the NRM as terminal context information. Figure 6.6 illustrates the various NRM and TRM functions defined in IEEE 1900.4; Table 6.1 summarizes the information exchanged between the various entities.

6.5.2.3 Information Model
Because IEEE 1900.4 targets heterogeneous wireless environments, the context and policy information used for decision-making must be agnostic to any specific CWN or terminal. As their solution, IEEE 1900.4 specifies an information model to

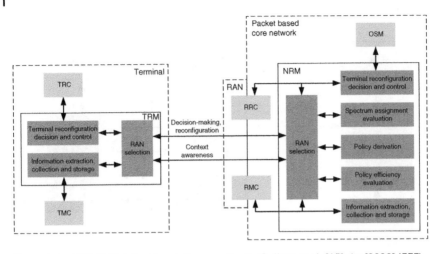

Figure 6.6 IEEE 1900.4 functional diagram. *Source:* Buljore et al. [15]. (© [2020] IEEE).

Table 6.1 IEEE 1900.4 interfaces.

NRM → TRM	• Radio resource selection policies • RAN context information • Terminal context information for other TRM	TRM → NRM	• Terminal context information for this TRM
NRM → RRC	RAN reconfiguration requests	RRC → NRM	RAN reconfiguration responses
NRM → RMC	• RAN context information requests	RMC → NRM	• RAN context information responses
NRM → OSM	• Information on spectrum assignments	OSM → NRM	• Spectrum assignment policies
TRM → TRC	• Terminal reconfiguration requests	TRC → TRM	• Terminal reconfiguration responses
TRM → TMC	• Terminal context information requests	TMC → TRM	• Terminal context information responses

Source: Buljore et al. [15].

support their functional requirements. Key tenants of the modeling approach include the following: an extensible form to accommodate future radio access technologies and allow for custom extensions to existing data models, an object-oriented approach, and exclusivity or consistency relationships between objects to determine conflicts.

Since the CWN and terminal are part of the larger IEEE 1900.4 system, the information model treats these as two sets of managed objects. In all, (4) categories of classes are specified:

- Common base class
- Policy classes
- CWN-related classes
- Terminal-related classes

The common base class abstracts properties supported by all objects. These include instance names and generic events such as instantiation and deletion.

The policy classes abstract specific events and associated policies. That is, an event which triggers policy evaluation, pre-conditions for specific actions, and the actions themselves.

The CWN-related classes abstract the operator and RANs, where the operator concept includes assigned channels, regulatory rules, and spectrum assignment policies. The RAN concept includes base-stations and cells.

The terminal-related classes abstract the user, application, device, and radio-resource selection policy concepts; examples include terminal profiles, capabilities, and measurements.

6.6 Learning Algorithms

We now discuss several learning algorithms and their recent application to intelligent radio. Rather than survey a significant number of learning algorithms, as was aptly done in [41, 47], here we have chosen a smaller set to explore a bit further in depth. These include artificial neural networks, Markov models, and reinforcement learning. We reiterate there is no single algorithm required for intelligent radio i.e. no *"one ring to rule them all!"* The algorithm chosen by any resource manager should be the one deemed optimal for a specific modeling or learning task. These obviously vary by application, so from an implementation point-of-view, there must be some notion of system requirements. However, from a life cycle point-of-view, it may be advantageous for the architecture to be agnostic to the application itself. The CRM framework attempts to do just that, as the core facilitates protocol stack access to both the Knowledge Base and Toolbox. The IEEE 1900.4 architecture is less agnostic as [2] specifies both a data model and the specific behavior of each component. This is certainly not an assumption of which architecture will perform better in terms of wireless coexistence. It is however, an observation that ease of implementation with respect to the selection of learning algorithms may be aided by the availability of additional requirements.

6.6.1 Artificial Neural Networks

Artificial neural networks (ANN) have become a foundational component of modern machine learning architectures. At their core, these are commonly used to approximate functions, which otherwise are prohibitively complex, or have no known closed-form solution at all. This is reflected in the many classification and regression problems solved using ANNs. Examples relevant to spectrum activity include wireless signal identification and signal or channel parameter estimation. However, many higher-level learning algorithms also benefit from some form of function approximation. As such, ANNs are often used as components within other learning algorithms. A prime example is deep reinforcement learning (DRL), where deep neural networks (DNN) are used to learn the Q-function. DRL has been applied to DSA and optimization applications where the resource manager learns to select an appropriate channel or waveform configuration through continued interaction with the environment. We elaborate on reinforcement learning later in this chapter.

ANNs are biological-inspired computational algorithms [48, 49] thought to mimic brain activity. Unlike a computer, which has a single processor, the brain has many processing units called **neurons**, operating in parallel. Each neuron is connected to $\sim 10^4$ other neurons via *synapses*, all processing information in parallel. To be clear, there is no assumption that a human can process the amount of information a computer can, or anywhere near as fast. However, we do know that even at an early age, humans demonstrate significant learning capacity, which computers have not yet been able to mimic.

Suggested neural network architecture is evolving rapidly. In fact, the many architectures published in journal articles and conference proceedings far outpace the material published in textbooks. These still provide valuable information regarding background material and basic concepts; for example, [49] is an excellent reference on deep learning. This is only to say that keeping abreast of the latest architectures, best practices, and suggested applications requires continual paper surveys, which may be true of many disciplines within machine learning and artificial intelligence. Here we review several families of architectures that have particular relevance to intelligent radio. These include the multi-layer perceptron (MLP), convolutional neural network (CNN), recurrent neural network (RNN), auto-encoder (AE), and adversarial auto-encoder (AAE). For a complete treatment on ANNs and associated background material, we refer you to references [48–50].

6.6.1.1 Multilayer Perceptron

ANNs process information in layers, with the simplest architecture consisting of two layers, an input and output as shown in the simple perception illustrated in Figure 6.7a. A simple perceptron is commonly labeled **dense** or **fully connected**.

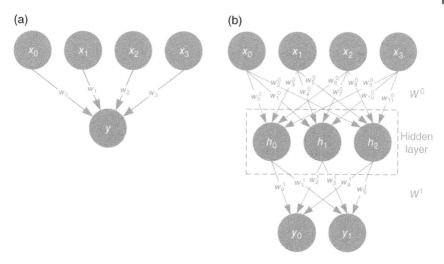

Figure 6.7 Simple perceptron (a) and multilayer perception (b).

Here the output is a function of the weights between the N input neurons and single output neuron as described in Eq. (6.1); b is an optional bias, and σ is an optional activation function. Activation functions are used to model highly nonlinear functions; commonly used functions include *sigmoid*, *rectified linear* (ReLU), *hyperbolic tangent* (tanh), and *softmax*.

$$y = f(x) = \sigma\left(\sum_{i=0}^{N} w_i x_i\right) + b \tag{6.1}$$

$$y = g(f(x)) = \sigma\left(W^1 \sigma\left(W^0 x + b^0\right) + b^1\right) \tag{6.2}$$

To increase the learning capacity, we can add additional layers to construct the multilayer perceptron (MLP) illustrated in Figure 6.7b. The one shown has one internal layer, and two output neurons. We call the internal layers **hidden** as we do not necessarily have visibility into their behavior. Following our earlier notion, we can express either of the outputs as Eq. (6.2). When the number of hidden layers exceeds two, the architecture is commonly labeled a **deep neural network**; hence the term **deep learning**. This definition is debatable, as some consider three to five hidden layers as deep, rather than two. What is more important is the change in learning goals represented by the additional depth. Shallow networks map input features to the desired output per the learned function approximation. In contrast, deep networks learn the features mapped to the output; that is, they learn a useful representation of the input data, along with the desired function approximation [49]. It does so by representing complex concepts as a hierarchy of simpler ones. For example, object recognition can be performed by feeding image pixels directly into a CNN. As the network is traversed in depth, edges illuminate contours and corners, which illuminate object parts, which illuminate objects. From this example, we see that deep learning eliminates both the time required to engineer relevant features from raw data, and the inductive bias discussed in [48]. Recalling the well-known *no free lunch* theorem [48], the increased capacity comes at the expense of network training. Deep networks are more difficult to train than shallower networks. Furthermore, there are practical limits to network depth as we shall soon see.

From Figure 6.7 we see that information flows forward through the network with respect to the desired output. As such, the MLP is considered a *feed-forward* network. The output is commonly labeled a **prediction**, whose error e, depends upon the specified loss function. For example, if designing a network to perform regression, the mean squared error (MSE), shown in Eq. (6.3), is typically chosen, where y and $\hat{y}$ are the target and prediction of the activated neuron. For a classification task, the cross-entropy function, shown Eq. (6.4), is typically chosen, where C represents the number of classes.

$$e = \frac{1}{2}(y - \hat{y})^2 \tag{6.3}$$

$$e = \sum_{i=1}^{C} -y_i \log(\hat{y}_i) \tag{6.4}$$

To train the network, we use an iterative process to continually update all network weights until the prediction accuracy is satisfactory. In contrast to the prediction process, information must flow backward, such that the total prediction error E is propagated back toward the input. This is commonly referred to as the **backpropagation algorithm** [49, 51]. Because of the non-linear behavior of each layer, specific weight adjustment is not proportional to prediction error. As such, we need to determine the **gradient** or relative change in prediction error with respect to each specific weight; for this, we utilize the chain rule for composite functions [52]. This is shown in Eq. (6.5) for the prediction layer of the MLP illustrated in Figure 6.7b. For this example, we assume a sigmoid activation function for all neurons, and the MSE loss function. Once the gradient is calculated, w_0^1 can be updated per Eq. (6.6), using a designer specified learning rate η. Although w_0^1 is shown as an example, w_1^1 thru w_5^1 are calculated in similar fashion.

$$\Delta w_0^1 = \frac{dE}{dw_0^1} = \frac{dE}{d\sigma_{y_0}} \frac{d\sigma_{y_0}}{dy_0} \frac{dy_0}{dw_0^1} = (-(\hat{y}_0 - y_0))(y_0(1 - y_0))(h_0) \tag{6.5}$$

$$w_0^1 = \eta \Delta w_0^1 \tag{6.6}$$

For the hidden layer, we use a similar process while recognizing that each weight now contributes to multiple prediction error terms. This is described in

Eqs. (6.7) and (6.8) for w_0^0 of the MLP shown in Figure 6.7b, where Eq. (6.8) represents the back-propagated error terms. Again, once the gradient is calculated, w_0^0 can be updated per Eq. (6.9), using a designer specified learning rate η. Although w_0^0 is shown as an example, w_1^0 thru w_{11}^0 are calculated in similar fashion.

$$\Delta w_0^0 = \frac{dE}{dw_0^0} = \frac{dE}{d\sigma_{h_0}}\frac{d\sigma_{h_0}}{dh_0}\frac{dh_0}{dw_0^0} = \left(\frac{dE}{d\sigma_{h_0}}\right)(h_0(1-h_0))(x_0) \tag{6.7}$$

$$\frac{dE}{d\sigma_{h_0}} = \frac{dE_0}{d\sigma_{h_0}} + \frac{dE_1}{d\sigma_{h_0}} = (-(\widehat{y_0}-y_0))(y_0(1-y_0))(w_0^1)$$
$$+ (-(\widehat{y_1}-y_1))(y_1(1-y_1))(w_1^1) \tag{6.8}$$

$$w_0^0 = \eta\Delta w_0^0 \tag{6.9}$$

The iterative process of updating the weights using gradients is known as **gradient descent** due to the idyllic convex shape of the error surface. In reality, the error surface likely has multiple local minima in addition to a global minimum. Regardless, updating the weights per the average gradient over the entire training set quickly becomes prohibitive in terms of training time and computational resources. The opposite approach is **stochastic gradient descent**, where the weights are updated per a single random example. A common compromise is **mini-batch gradient descent**, where the weights are updated per the average gradient of a subset of the training set. When building a model within a particular development environment, the choice of how many examples to include per gradient update is typically specified as the **batch size**.

For a fixed-size training set, one pass through its entirety is considered an **epoch**. It is common for a network to be trained over hundreds or thousands of epochs, and therefore an even greater number batches depending upon the batch and epoch sizes chosen.

6.6.1.2 Convolutional Neural Networks

Perhaps no other architecture has contributed more to the recent popularity of deep learning than the CNN. Its versatility and inference efficiency as compared to the MLP have inspired many designers to rethink existing approaches to function approximation in favor of deep learning. It is easy to see why when considering three specific traits: sparse connectivity, parameter sharing, and equivariant representations [49]. By definition, a CNN is an MLP with one or more layers implemented using **convolution** rather than matrix multiplication. Here, the term *convolution* is equivalent to the two-dimensional cross-correlation described in Eq. (6.10), where D is the data array being processed, and K is the kernel. This operation is often expressed in two dimensions given its popular application to image processing; however, this can be reduced to a single dimension through removal of the m index, or expanded to three or greater dimensions if desired.

Also, this deviates from our classical definition of convolution only in that the kernel is not flipped, which necessarily assumes the kernel is real valued.

$$S(i,j) = (D*K)(i,j) = \sum_m \sum_n D(i + m, j + n)K(m, n) \tag{6.10}$$

Equation (6.10) is illustrated in Figure 6.8 for a stride of one across both i, j indices; when $m < i$, we see that each output is not dependent upon every input. This is in stark contrast to the MLP in Figure 6.7b, where each output neuron in the hidden layer is dependent upon every input value x. This **sparse connectivity** facilitates the learning of meaningful features with much fewer operations as compared to the MLP. We also see that the same kernel is applied to the entire input array, rather than learning a unique weight for each input value. This **parameter sharing** significantly reduces architecture memory requirements as compared to the MLP. These traits combined suggest a CNN can be much more efficiently inferred, and within a much smaller memory footprint than a MLP with similar functionality. Parameter sharing also means the CNN is equivariant to translation. That is, any shift in the input causes a similar shift in the output feature map. This is extremely useful when the position of specific information within an input stream is variable, e.g. the location of an object within an image or time of arrival or phase of a signal.

Earlier we mentioned that there are practical limits to network depth. In [53], He et al. demonstrate a performance saturation above a certain DNN depth. They reason that multiple nonlinear layers have difficulty approximating an identity mapping. Their solution is to implement a periodic *skip* connection, as shown in Figure 6.9, such that the stacked layers learn a **residual** mapping rather than an identity mapping; hence the term **ResNet**. Because of the skip connection, the stacked layers learn $H(x) - x$ rather than $H(x)$, which the authors demonstrate to be an easier task. The forced identity mapping then produces the desired approximation of $H(x)$ prior to the second activation function. Given its demonstrated success, ResNets are now widely used as a baseline for many CNN architectures.

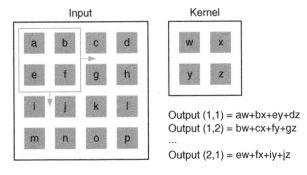

Figure 6.8 2D convolution operation.

Figure 6.9 Residual block.

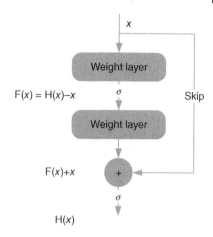

$F(x) = H(x) - x$

$F(x) + x$

$H(x)$

6.6.1.3 Recurrent Neural Networks

So far, we have implicitly assumed that the input sequence is comprised of *independently and identically distributed (iid)* random variables. For a large class of input values, this is not the case. For example, hourly rainfall measurements, daily currency exchange rates, and the sequence of characters in a word or sentence are clearly dependent. The feed-forward architectures discussed thus far are generally incapable of recognizing this dependency, as they have no native means to retain information during inference. As such, they perform poorly at sequence modeling and event prediction without significant modification. We now explore the recurrent neural network (RNN), which contains specific elements to *remember* interim values over arbitrary time intervals. Early architectures suffered from short-term memory due to vanishing gradients during back propagation [54, 55]; this issue has since been mitigated through the use of **Long Short-Term Memory (LSTM)**. As these are now common library elements within most ANN development environments, we deem these basic RNN building blocks. For full historical context, we refer the reader to [49, 55].

RNNs model vector sequences through selective memory; that is, they learn to remember what is important, and discard what is not [49, 56]. The LSTM-based RNN does so via a cell state, internally managed by three gates: a forget gate, input gate, and output gate. This is illustrated in Figure 6.10a where the explicit feedback connections for a network slice with input x_t are shown for both the cell state C_t, and output h_t. Figure 6.10b removes the feedback connections by unrolling the network slice over three time steps; it should be clear that the functionality of the two versions are the same.

The architecture of the modern LSTM is illustrated in Figure 6.11. Starting with the cell state from the previous iteration, C_{t-1} is scaled by the forget gate to discard none, some, or all of the cell state. f_t is a function of h_{t-1}, x_t, and their respective

(a)

(b)

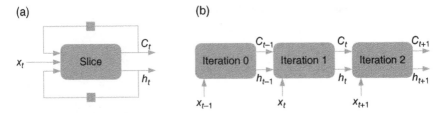

Figure 6.10 LSTM-based RNN.

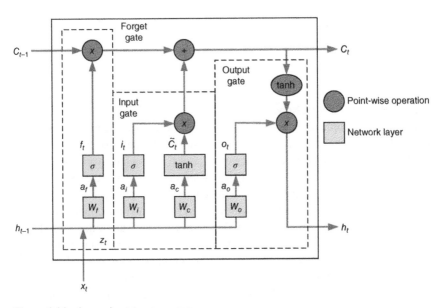

Figure 6.11 Long short-term memory.

weights $\boldsymbol{W}_f$ as shown in Eq. (6.11); $\boldsymbol{z}_t = (\boldsymbol{h}_{t-1}, \boldsymbol{x}_t)$, b_f is a bias value, and σ the sigmoid activation function. Next, the scaled cell state is updated per the input gate. i_t and $\widetilde{\boldsymbol{C}}_t$ are functions of $\boldsymbol{h}_{t-1}$, $\boldsymbol{x}_t$, and their respective weights $\boldsymbol{W}_i$ and $\boldsymbol{W}_C$ as shown in Eqs. (6.12) and (6.13). i_t is activated using a sigmoid, and $\widetilde{\boldsymbol{C}}_t$ using tanh. The current cell state $\boldsymbol{C}_t$ is then calculated per Eq. (6.14), and the output $\boldsymbol{h}_t$ determined per the output gate. $\boldsymbol{C}_t$ is first activated using tanh, and then scaled per Eq. (6.16); o_t is a function of both $\boldsymbol{h}_{t-1}$ and $\boldsymbol{x}_t$ and their respective weights $\boldsymbol{W}_o$ as shown in Eq. (6.15).

$$f_t = \sigma\left(\boldsymbol{W}_f \boldsymbol{z}_t + b_f\right) \tag{6.11}$$

$$i_t = \sigma\left(\boldsymbol{W}_i \boldsymbol{z}_t + b_i\right) \tag{6.12}$$

$$\widetilde{C}_t = \tanh{(W_C z_t + b_C)} \tag{6.13}$$

$$C_t = f_t C_{t-1} + i_t \widetilde{C}_t \tag{6.14}$$

$$o_t = \sigma(W_o z_t + b_o) \tag{6.15}$$

$$h_t = o_t \tanh{(C_t)} \tag{6.16}$$

The various LSTM weight vectors are learned in a similar fashion to the MLP and CNN [56]. The gradients are calculated by applying backpropagation to the unrolled network shown in Figure 6.10b; that is, the gradients accumulate over all time steps T. For example, ΔW_o is calculated for a single time step in Eq. (6.17). We again assume the MSE loss function for the output h_t. Equation (6.17) is illustrated in Figure 6.12, where $\frac{dE}{dh_t}$ is propagated backwards toward W_o. W_o is then updated per Eq. (6.18) after averaging ΔW_o over the entire mini batch. Although W_o is shown as an example, W_f, W_i, and, W_C are calculated in similar fashion.

$$\Delta W_o = \frac{dE}{dW_o} = \frac{dE}{dh_t}\frac{dh_t}{do_t}\frac{do_t}{da_o}\frac{da_o}{dW_o} = -\left(\left(\widehat{h}_t - h_t\right)\right)(\tanh{(C_t)})(o_t(1 - o_t))(z_t) \tag{6.17}$$

$$W_o = \eta \sum_T \Delta W_o \tag{6.18}$$

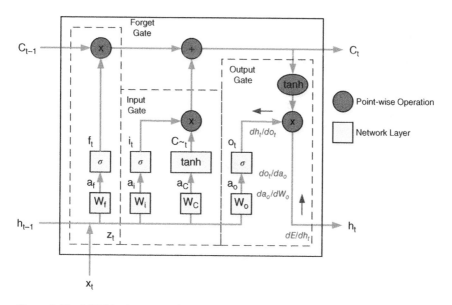

Figure 6.12 LSTM back propagation.

6.6.1.4 Generative Models

Thus far, we have discussed MLP, CNN, and RNN architectures used for function approximation. We next explore architectures used to learn data distributions. Why is this useful? Because we can use these to supplement data sets, generate samples with impairments removed, generate samples with specific attributes, or detect anomalies. All these tasks generate new data points with some variation; hence the term **generative model**. Each, however, requires a model that has learned the underlying probability distribution of the training data; otherwise, the samples generated become dissimilar to the point of being inadequate for the given task. In this section we investigate two specific architectures currently in widespread use, the Variational Autoencoder (VAE), and Adversarial Autoencoder (AAE). These are both built upon the function approximations we have already discussed, and are also trained using some form of gradient descent.

A VAE is best understood when compared to a standard Autoencoder (AE) [49] shown in Figure 6.13a. It is composed of two functions: an encoder f, and decoder g. The encoder is trained to map the input x to a latent representation or code h; ideally, this code represents only the *important* features of x, which allows it to have a lower dimensionality. The decoder is trained to map the code h to the output r, which is similar but not necessarily equal to x. On the surface, there is no benefit in learning the identity function $x = r$ even though the AE is trained to do just that. However, what is of great benefit is for the AE to learn an ideal representation of x i.e. code h, such that any impairments are removed or missing information replaced during the reconstruction of x i.e. output r.

Both the encoder and decoder are comprised of combinations of the MLP, CNN, and RNN architectures discussed earlier; the exact combination is application dependent. For CNN decoder architectures, construction of r from h is primarily

(a)   (b)

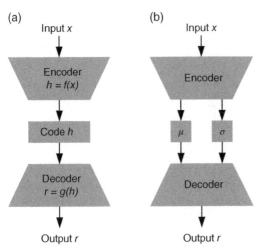

Figure 6.13 (a) Autoencoder and (b) variational autoencoder. *Source:* Based on Goodfellow et al. [49].

performed using one of two methods [57]. The first is some form of *upsampling* to produce the required output shape, including repeated input values, localized interpolation, and multiple zero-stuffing strategies. The second method is *transpose convolution*, in which the input from the preceding layer is first zero padded per the required output shape, and then the convolution operation, illustrated in Figure 6.8, is executed. Here the stride is indicative of the level of zero padding, rather than the stride of the kernel. The encoder/decoder pair is trained together using a loss function which penalizes any dissimilarity between r and x; the difference is commonly labeled *reconstruction error*. Assuming MSE, the reconstruction loss is described in Eq. (6.19). What should be clear is that given enough network capacity, the reconstruction error will be driven to zero, and the network will have essentially learned the identify function. As such, finding the balance between capacity and the desired output as a function of reconstruction loss is the primary design challenge.

$$L(x, g(\,f(x))) = \frac{1}{2}(x - r)^2 \tag{6.19}$$

The main drawback of the AE is that the latent space may be discontinuous, meaning that given a random input, the encoder may generate an h value the decoder is not familiar with. The decoder will then generate an unrealistic output rather than the desired variation. The VAE mitigates this by learning the mean μ and standard deviation σ of the distribution of x [58, 59]. As such, the decoder learns only to generate values governed by $N(\mu, \sigma^2)$. This is illustrated in Figure 6.13b.

The VAE latent representation presents two distinct differences with respect to network training. First, the encoder/decoder pair are trained together similar to the AE; however, what is passed on to the decoder is a random sample from the distribution $N(\mu, \sigma^2)$. The stochastic nature of the sampling process forces the decoder to learn that nearby points are members of the same class, eliminating any discontinuities within the latent space. Second, the VAE forces all classes close together by introducing a regularization term in the loss function. The regularization term is described in Eq. (6.20), where the regularization is the Kullback-Liebler (KL) divergence described in Eq. (6.21).

$$L(x, g(\,f(x))) = \frac{1}{2}(x - r)^2 - KL[N(\mu_x, \sigma_x), N(0, 1)] \tag{6.20}$$

$$KL[N(\mu_x, \sigma_x), N(0, 1)] = \mu_x^2 + \sigma_x^2 - \log(\sigma_x) - 1 \tag{6.21}$$

Clearly, the VAE is only able to model Gaussian distributions given this form of regularization. The AAE architecture expands the latent space modeling capability to arbitrary data distributions [60]. It does so by replacing the KL divergence penalty shown in Eq. (6.20) with the adversarial training procedure described in [61]. Particularly, the encoder is trained to produce a posterior distribution $q(z)$

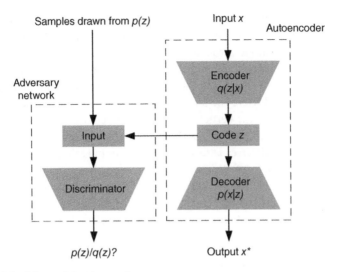

Figure 6.14 Adversarial autoencoder.

matching a prior distribution $p(z)$. The AAE is illustrated in Figure 6.14, and resembles an AE with a second network attached to the encoder output. $q(z)$ is described in Eq. (6.22), where $q(z\,|\,x)$ is the encoding distribution, and $p_d(x)$ is the input data distribution.

$$q(\mathbf{z}) = \int q(\mathbf{z}\,|\,\mathbf{x})p_d(\mathbf{x})d\mathbf{x} \qquad (6.22)$$

Assuming mini-batch gradient descent, AAE training is performed in two steps. The first is the *reconstruction phase*, where similar to the AE and VAE architectures, the AE updates the encoder and the decoder to minimize the reconstruction error. The second is the *regularization phase*, which itself is a two-step process. The AAE first updates the discriminator to discern between positive samples drawn from $p(z)$, and negative samples z from the encoder. It then updates the adversarial generator (AE encoder) to further confuse the discriminator such that it cannot discern between $p(z)$ and $q(z)$. Once trained, the AE decoder defines a generative model that maps the imposed prior $p(z)$ to the output data distribution. Also, as is discussed in [60], the AAE architecture can be expanded to incorporate categorical labels for both supervised and semi-supervised learning.

6.6.1.5 Example ANN Applications

In [62], Chew et al. investigate the use of deep learning to detect signal activity in the presence of both noise and interferers. Their motivation is the difficult task of weak signal detection in congested spectrum. Their approach is to perform binary classification on images (spectrograms) created from overlapping FFT blocks. Using transfer learning, they were able to retrain AlexNet for signal detection tasks using minimal amounts of training data. This included training data of different signal-to-noise ratios (SNR) and random interference. The result was a constant false alarm rate (CFAR) signal detection capability that did not require distinct noise estimation to achieve the desired CFAR operation.

In [63], Liu et al. investigate the use of deep learning to classify wireless signal modulation for cognitive radio applications. They argue that traditional methods of signal identification require a priori knowledge of both the signal in question and wireless channel. Furthermore, any a priori channel state information quickly loses accuracy as conditions change. Their approach is to train multiple DNN architectures to classify wireless signal modulation in the presence of synchronization errors and channel impairments. Architectures tested include residual networks (ResNet), densely connected networks (DesNet), and convolutional LSTM networks (CLDNN). Their dataset is comprised of 11 modulation types, with a block size of 128 baseband (complex) samples.

In [64], O'Shea et al. investigate the use of deep learning to estimate common synchronization parameters for cognitive radio applications. More specifically, they treat carrier frequency and clock offset estimation as regression problems. Similar to the motivation espoused in [63], they reason that analytic estimators are optimized for AWGN only, and perform poorly under channel and hardware impairments. Furthermore, analytic estimators require more computational resources than a DNN optimized for similar performance. Their approach is also to train multiple DNN architectures to estimate carrier frequency and timing error in the presence of channel impairments. Their dataset is limited to quadrature phase shift keying (QPSK) bursts of random payloads, with blocks sizes ranging from 32 to 1024 baseband (complex) samples.

In [65], Rajendran et al. present an architecture for spectral anomaly detection based upon an AAE. They motivate this through a discussion on the increased heterogeneous network traffic caused by the diversity of newer wireless protocols. This tends to decrease both the accuracy and efficiency of manual anomaly detection for management and regulatory purposes. Their solution is to train an AAE with multiple discriminators to execute the following tasks: reconstruct the given power spectral density (PSD), calculate the probability that the PSD comes from a known distribution, and predict the class of signal activity represented within the PSD. As such, an anomaly score is a combination of three metrics: reconstruction loss, distribution probability, and classification accuracy. The AAE is trained on examples of 6-consecutive 64-point FFTs, which essentially is a spectrogram of a selected frequency band for given block of time.

6.6.2 Markov Models

As was mentioned earlier, we often assume that observations are i.i.d. random variables. While this greatly simplifies calculation of the probability of occurrence, it is incorrect when successive observations are in fact dependent. As was demonstrated in our discussion on RNNs, dependent variables can be characterized using some form of parametric model. If we assume the underlying process to be comprised of distinct states, it can be characterized using a **Markov model** [48, 50]. This type of model is applicable to much of what is observed in the spectrum. Given little, or no a priori information about a given geographical area, spectrum activity is indeed a random process. However, the activity observed is dependent upon many underlying processes which can be characterized using discrete states. Examples include primary and secondary user activity, any underlying access protocols, and the channel state between nodes. The motivation being that once we derive a parametric model, we can then predict certain events and situations, which increase our ability to coexist with other spectrum users.

6.6.2.1 Observable States

Consider the Markov model shown in Figure 6.15. At any instant t, the system is in one of three states S, described by the state transition probability matrix α. The only exception is the initial state, described by the initialization probability vector π. Because the state at time $t + 1$ depends on the state at time t only, this is considered a first-order model. Here, these states are also observable, and each observation q_t represents the state of the system at time t. Given α, π we can calculate the probability of a sequence of observations O over time period T as the following:

$$P(O = Q \mid \alpha, \pi) = P(q_1) \prod_{t=2}^{T} P(q_t \mid q_{t-1}) = \pi_{q_1} \alpha_{q_1 q_2} \ldots \alpha_{q_{T-1} q_T} \tag{6.23}$$

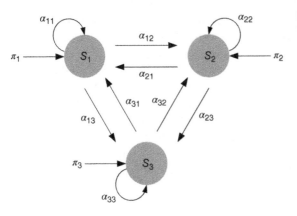

Figure 6.15 Markov model with three states.

If α, π are not given, these can be estimated from a sequence of observations by calculating the probability of each random variable. That is, for each variable, dividing the sum of desired outcomes by the total number of possible outcomes. This is shown in the following equations assuming access to K sequences of length T.

$$\pi_i^* = \frac{\sum_k \left(q_1^k = S_i\right)}{K} \tag{6.24}$$

$$\alpha_{ij}^* = \frac{\sum_k \sum_{t=1}^{T-1} \left(q_t^k = S_i \text{ and } q_{t+1}^k = S_j\right)}{\sum_k \sum_{t=1}^{T-1} \left(q_t^k = S_i\right)} \tag{6.25}$$

6.6.2.2 Hidden Markov Model

When the states are not observable, a stateful process can be modeled as a **Hidden Markov Model** (HMM) [48, 50]. Here we observe a probabilistic function of each state v_m, rather than the explicit state itself q_t. This is illustrated in Figure 6.16 for the three state example shown in Figure 6.15, but with the states hidden and two possible observations per state. The probability of each observation $b_j(m)$ is expressed using the following relation:

$$b_j(m) \equiv P\left(O_t = v_m \mid q_t = S_j\right) \tag{6.26}$$

Note that Eq. (6.26) expresses two sources of uncertainty: the state transitions themselves and the observations within each state. To elaborate, when the states are observable the observation itself is indicative of the state. For an HMM we must also account for the probability that a given observation is indicative of a particular state. As such, HMM's are defined by three parameters: the state transition

Figure 6.16 Hidden Markov model with three states and two possible observations. *Source:* Based on Alpaydin [48] and Bishop [50].

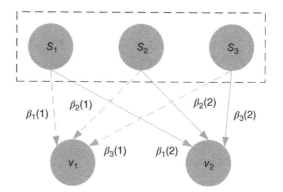

probability matrix α, the observation probability vector β, and the initial state probability vector π. These are often expressed as the tuple $\lambda = (\alpha, \beta, \pi)$.

In practice, we are often provided only observations, and make the decision to model the underlying process as an HMM. The benefit being that we can then evaluate the probability of any sequence of observations or determine the sequence of states most likely to have produced a given sequence of observations. In either case, our first task is to learn a model λ^* which maximizes $P(O \mid \lambda)$; for this, we utilize a specific expectation–maximization (EM) procedure called the Baum-Welch algorithm. This is an iterative approach to learning λ^*. While training, it converges to a *locally* optimal solution, e.g. a local extrema. It should be noted that the solution is not guaranteed to be globally optimal, but should be *close enough* and also stable.

We now define the set of variables required to execute the Baum-Welch algorithm.

$$\theta_t(j) = P\big(O_1...O_t, q_t = S_j \mid \lambda\big) = \left[\sum_{i=1}^{N} \theta_{t-1}(i)\alpha_{ij}\right]\beta_j(O_t) \tag{6.27}$$

$$\phi_t(i) = P(O_{t+1}...O_T \mid q_t = S_i, \lambda) = \sum_{j=1}^{N} \alpha_{ij}\beta_j(O_{t+1})\phi_{t+1}(j) \tag{6.28}$$

$$\xi_t(i,j) = P\big(q_t = S_i, q_{t+1} = S_j \mid O, \lambda\big) = \frac{\theta_t(i)\alpha_{ij}\beta_j(O_{t+1})\phi_{t+1}(j)}{\sum\limits_{j=1}^{N} \theta_t(j)\varphi_t(j)} \tag{6.29}$$

$$\gamma_t(i) = P(q_t = S_i \mid O, \lambda) = \sum_{j=1}^{N} \xi_t(i,j) \tag{6.30}$$

The forward looking variable $\theta_t(j)$ describes the probability of arriving in state S_j, regardless of path, given all previous observations. The backward looking variable $\varphi_t(i)$ describes the probability of starting in state S_i, regardless of path, given all ensuing observations. $\xi_t(i, j)$ describes the probability of transitioning from state S_i to state S_j at time t, given all observations. And finally, $\gamma_t(i)$ describes the probability of being in state S_i at time t, given all observations.

We start Baum-Welch by initializing $\lambda = (\alpha, \beta, \pi)$ with either random numbers or values based on some a priori information. We then iterate over the two-step EM process. During the E-step, we update $\theta_t(j)$, $\varphi_t(i)$, $\xi_t(i, j)$, and $\gamma_t(i)$ per Eqs. (6.27)–(6.30) Next, during the M-step, we update $\lambda^* = (\alpha^*, \beta^*, \pi^*)$ per Eqs. (6.31)–(6.33). This process repeats until $P(O \mid \lambda)$, the denominator of Eq. (6.29), is maximized, i.e. exhibits minimal change without decreasing.

$$\pi^*(i) = \gamma_1(i) \tag{6.31}$$

$$\alpha^*_{ij} = \frac{\sum\limits_{t=1}^{T-1} \xi_t(i,j)}{\sum\limits_{t=1}^{T-1} \gamma_t(i)} \tag{6.32}$$

$$\beta^*_j(m) = \frac{\sum\limits_{t=1}^{T-1} \gamma_t(i)s.t.O_t = v_m}{\sum\limits_{t=1}^{T-1} \gamma_t(i)} \tag{6.33}$$

6.6.2.3 Gaussian Distributions

In our HMM example above, we assumed a set of discrete observations such that the observation probabilities are discrete conditional distributions as shown in Eq. (6.33). If the observations are in fact continuous, that is $O_t \in \mathcal{R}$, we can rewrite $P(O_t = v_m \mid q_t = S_j)$ in terms of a Gaussian distribution. As an example, we assume scalar observations and a normal distribution, and rewrite Eq. (6.26) as Eq. (6.34). This implies that while in state S_j, the observations are drawn from a normal distribution with mean μ_j and variance σ^2_j.

$$b_j(m) = P\big(O_t \mid q_t = S_j, \lambda\big) \sim \mathcal{N}\big(\mu_j, \sigma^2_j\big) \tag{6.34}$$

Here we do not develop the continuous time EM procedure used to learn the model. We do however suggest that in some cases, modeling the observations using a Gaussian or mixture of Gaussians may in fact more accurately model the various metrics being observed.

6.6.2.4 Example Markov Model Applications

In [66], Akbar and Tranter use an HMM to predict PU channel activity prior to SU channel access. Their premise is that predicting PU activity results in lower overall signal-to-interference ratios (SIR) for the PU. The reason being that each SU looks for PU activity at the start of a given timeslot, so there is always potential for collision; carrier sense multiple access (CSMA) is used for comparison purposes. In their approach, each PU channel is modeled an HMM. Based on its immediate history, if the probability of the PU being active is greater than not, the SU vacates that channel prior to the timeslot in question; hence, any collision is avoided. For simulation purposes, PU activity was modeled using a Poisson distribution with a timeslot occupancy rate ≥ 0.5.

In [67], Senthilkumar and Geetha employ an HMM within a larger framework for PU/SU channel selection. Here the channel bandwidth requirements are matched with actual channel conditions to select the optimal channel. An

HMM is modified such that the output probabilities represent the fitness of each channel within the network. Accordingly, these are a function of available bandwidth, noise power, and the current state estimate. If the probability exceeds a threshold, that channel is allocated to a PU/SU for use in the next time slot. Similar to [66], this approach avoids the latency of channel sensing and estimation from within a given timeslot.

In [68], Li and Xiong develop a Markov Model Bank (MMB) to analyze large cognitive radio networks comprised of both PUs and SUs. Their motivation is derived from the assumption that larger networks are heterogeneous in nature, meaning each radio may have different capabilities and policies which guide their respective spectrum access requirements and decision-making. Given this dissimilarity between SUs and channels, characterizing overall SU throughput becomes prohibitively complex. Rather than model all SUs as a single entity, their solution is to model each SU as a Markov model comprised of $3K + 1$ states. Each of the K channels is represented by three states: sensing, transmission, and waiting; the one common state is that of switching between channels. This approach provides spatial, channel, and user de-correlation, which greatly simplifies throughput calculations.

6.6.3 Reinforcement Learning

Now armed with methods for function approximation and time series analysis, the next step is to utilize this information for decision-making. That is, given some information about the environment, we would like to act in a manner which furthers our own interests. For purposes of intelligent radio, that could be any action in support of the many use-cases described earlier in this Chapter. Examples include frequency agility, waveform reconfiguration, or to make no changes at all. Recall we also developed a cognitive cycle for intelligent radio operation. As shown in Figure 6.3, this is an iterative process, requiring a series of decisions informed by both current and past observations within a given environment. Exactly how to make these decisions is the focus of this section.

We consider the spectrum to be a stochastic environment in the sense that we are able to control our actions within a given environment, but not the environment itself. This makes sense as many actors utilize the spectrum, each with their own individual interests of which we have no knowledge. Furthermore, wireless channels themselves vary continuously with the many physical aspects of each location. We can model these using various gain profiles and distributions, but at any instant, we can at best produce a time-sensitive estimate of its characteristics. The result is an inability to map actions to specific outcomes a priori. If this were true, we could treat decision making as a supervised learning task, and never need to *learn* how to operate within a given environment. Unfortunately, the opposite is true, where an intelligent radio must make a series of decisions informed only by its observations to date. We tackle this problem using Reinforcement Learning (RL). That is, we learn to make *good* decisions through continued interaction with the environment [69]. Each decision results in actions providing observable effects on the environment, and we use these to inform ensuing decisions with respect to our goals. Our overarching assumption is that after a period of time, we will have gained enough experience to effectively act within a given environment to achieve our goals.

6.6.3.1 Learning by Experience

RL problems are described by a system comprising an agent and its environment. The agent operates within this environment while trying to achieve its goal. The environment produces information indicative of its **state**, which is used by the agent to select its next **action**. Each action induces a state change and some form of **reward**, both observed by the agent. This concept is illustrated in Figure 6.17, where each time step t represents a *state → action → reward* sequence; this is often expressed as the tuple (s_t, a_t, r_t), and labeled an *experience*. Repeated experiences require an agent to make a series of decisions, which ultimately *reinforce* good actions, and de-emphasize others.

The agent's **policy** maps environment states to specific actions. It does so based upon an *objective*, which represents the aggregate of all previous rewards. As such, a *good* policy is one that maximizes the objective, which is learned only through continued interaction with the environment.

In learning the policy, the environment state transition probabilities must be accurately modeled. Consider the generic transition function shown in Eq. (6.35). For an **episode** of t time steps, the transition probabilities for state s_{t+1} depends upon all previous state-action pairs. For large t, modeling becomes intractable given the vast combination of effects. However, this is greatly simplified if assuming the Markov property. That is, s_{t+1} depends only upon (s_t, a_t) as shown in Eq. (6.36). Furthermore, we can now model sequential decision-making as a **Markov Decision Process** (MDP) [69, 70] defined by the tuple $(S, A, P(.), R(.))$, where S is the set of states defining the environment, A is the set of possible

Figure 6.17 Reinforcement learning control loop.

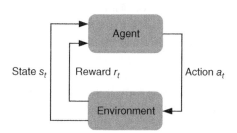

actions by the agent, $P(s_{t+1} \mid s_t, r_t)$ is the environment state transition function, and $R(s_t, a_t, s_{t+1})$ is the environment reward function. We reiterate that for RL problems, the agent has no a priori knowledge of either $P(.)$ or $R(.)$; these are learned through experiences (s_t, a_t, r_t).

$$s_{t+1} \sim P(s_{t+1} \mid (s_0, a_0), (s_1, a_1), ..., (s_t, a_t)) \tag{6.35}$$

$$s_{t+1} \sim P(s_{t+1} \mid s_t, a_t) \tag{6.36}$$

Defining the objective in terms of rewards, we first introduce the *return* $R(\tau)$ shown in Eq. (6.37). τ indicates a specific *trajectory* or sequence of experiences over an *episode* $t = 0 : T$; the return is the discounted sum of rewards over t. The discount factor γ is a design parameter $\in [0, 1]$, which scales future rewards. As $\gamma \to 0$, the return becomes shortsighted and scales future rewards less; as $\gamma \to 1$, the return becomes farsighted and scales future rewards more. The objective is then simply the expectation of the return, as shown in Eq. (6.38).

$$R(\tau) = \sum_{t=0}^{T} \gamma^t r_t \tag{6.37}$$

$$J(\tau) = \mathbb{E}_\tau[R(\tau)] = \mathbb{E}_\tau \left[\sum_{t=0}^{T} \gamma^t r_t \right] \tag{6.38}$$

As is shown in Figure 6.17, an agent can learn a policy π which maps environment states to specific actions. This is formally described as $a \sim \pi(s)$, which indicates action a is sampled from policy π given state s. Alternatively, an agent could learn one of two other functions. The first is a model of the environment, which is essentially the transition function described in Eq. (6.36). This allows the agent to predict the effects of various actions, and select the one which maximizes the objective at each time step t. The other is a *value* function, which provides information about the objective itself. That is, it provides a quantitative measure of states and actions in terms of the anticipated return $R(\tau)$. For the rest of this section, we will focus on policy functions, value functions, and a combination of the two known as **Advantage Actor-Critic** (A2C).

6.6.3.2 Policy Gradients

As stated earlier, the goal is to learn a policy π, which maximizes the objective. We first restate this in terms of a trajectory generated from a specific policy with parameters θ. This is shown in Eqs. (6.39) and (6.40) for the return and objective functions. Note the return only discounts rewards beyond t', and the objective is calculated over many trajectories sampled from π_θ. Also, recall earlier in this chapter we discussed a powerful function approximation method, namely artificial neural networks. We therefore assume the optimal policy can be learned using

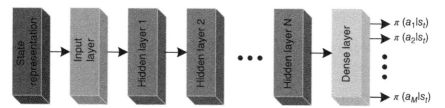

Figure 6.18 Deep policy network.

a deep policy network (DPN). This is illustrated in Figure 6.18, where the DPN is comprised of N hidden layers, and evaluates the probability of M possible actions per state.

$$R_t(\tau) = \sum_{t'=t}^{T} \gamma^{t'-t} r_t'$$

(6.39)

$$J(\pi_\theta) = \mathbb{E}_{\tau \sim \pi_\theta}[R(\tau)] = \mathbb{E}_{\tau \sim \pi_\theta}\left[\sum_{t=0}^{T} \gamma^t r_t\right]$$

(6.40)

To learn the parameters that maximize the objective, we utilize gradient ascent as shown in Eq. (6.41). Similar to our discussion on the Multilayer Perceptron, the gradient points in the direction of steepest ascent, and η is a user specified learning rate. The gradient of interest here, the **policy gradient**, is defined in Eq. (6.42); we refer the interested reader to references [69, 71] for extensive derivations of the policy gradient. What is important, is that $\pi_\theta(a_t \mid s_t)$ represents the probability of the action $a_t \sim \pi_\theta(s_t)$ taken at time t. Furthermore, the gradient of the objective is equivalent to the expected sum of the gradients of the log-probabilities of a_t multiplied by the corresponding returns. This indicates how to accumulate each parameter update over each trajectory sampled from π_θ.

$$\theta \rightarrow \theta + \eta \nabla_\theta J(\pi_0)$$

(6.41)

$$\nabla_\theta J(\pi_\theta) = \mathbb{E}_{\tau \sim \pi_\theta}\left[\sum_{t=0}^{T} R_t(\tau) \nabla_\theta \log\left(\pi_\theta(a_t \mid s_t)\right)\right]$$

(6.42)

The REINFORCE algorithm [69, 72] is commonly used to estimate the policy gradient. It does so through Monte Carlo sampling of single trajectories over T time steps per episode. Because its parameter update depends upon the current policy, REINFORCE is considered an **on-policy** algorithm. As such, each trajectory recorded contributes to a single parameter update only, and is discarded afterwards. This is not the case when learning a value function, as we will soon see.

6.6.3.3 Q-Learning

There are two forms of the value function, $V^\pi(s)$ and $Q^\pi(s, a)$, as shown in Eqs. (6.43) and (6.44). $V^\pi(s)$ measures the expected return, according to policy π, from state s_0 of trajectory τ to the end of the episode. Similarly, $Q^\pi(s, a)$ measures the expected return, according to policy π, from state-action pair (s_0, a_0) of trajectory τ to the end of the episode. We narrow our focus by further exploring $Q^\pi(s, a)$ or **Q-Learning** as is commonly referred to. Why? First, converting $Q^\pi(s, a)$ into a policy is easier than $V^\pi(s)$ as it contains more information; namely state-action pairs instead of actions only. Second, as was the case for policy gradients, we can utilize artificial neural networks to learn $Q^\pi(s, a)$ in the form of **Deep-Q Networks** (DQN). This was first attempted by Minh et al. [73] in teaching an agent to play video games. The input was raw pixels, and the reward the game score. Our application of intelligent radio is obviously very different; however, their training methods are employed to this day.

$$V^\pi(s) = \mathbb{E}_{s_0, \tau \sim \pi}\left[\sum_{t=0}^{T} \gamma^t r_t\right] \tag{6.43}$$

$$Q^\pi(s, a) = \mathbb{E}_{s_0, a_0, \tau \sim \pi}\left[\sum_{t=0}^{T} \gamma^t r_t\right] \tag{6.44}$$

6.6.3.4 Deep Q-Networks

DQN's ingest a representation of the current state s, and output $Q^\pi(s, a)$ for each possible action a. This is shown in Figure 6.19, where the DQN is comprised of N hidden layers, and evaluates M possible actions. As such, it must learn $Q^*(s, a)$, which is the optimal Q-function as described in Eq. (6.45). That is, $Q^*(s, a)$ corresponds to the optimal policy π^*. DQN is an **off-policy** algorithm, meaning the function being learned is independent of the policy used to interact with the environment during training. Why does this matter? Because, we can learn something from each individual experience rather than requiring an entire episode to update $Q^\pi(s, a)$. This facilitates an iterative approach to learning $Q^*(s, a)$, which is much

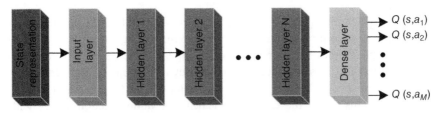

Figure 6.19 Deep Q-network.

more sample efficient than an *on-policy* algorithm. The update is accomplished via the Bellman equation shown in Eq. (6.46), which represents the value of state-action pair (s', a') as a function of the previous state-action pair (s, a) and the associated reward r. We can use this to generate the DQN target values shown in Eq. (6.47), which are required to update the neural network parameters as described earlier in this chapter.

$$Q^*(s, a) = \max_\pi Q^\pi(s, a) = Q^{\pi^*}(s, a) \tag{6.45}$$

$$Q^\pi(s, a) \approx r + \gamma \max_{a'} Q^\pi(s', a') \tag{6.46}$$

$$Q^\pi_{\text{tar}}(s, a) = r + \gamma \max_{a'} Q^\pi(s', a') \tag{6.47}$$

Recall the MSE loss function described in Eq. (6.3), and the gradient descent algorithms introduced in the Multilayer Perceptron discussion. For a DQN with parameters θ, the loss function is shown in Eq. (6.48), and the parameter update shown in Eq. (6.49). The experiences used to calculate $Q^\pi(s, a)$ and $Q^\pi_{\text{tar}}(s, a)$, are stored while interacting with the environment using the current policy. This suggests a two-step iterative training process [69, 73], whereby the agent first interacts with the environment for some finite time period using the current policy. The results of these interactions, the experiences, are stored in memory as shown in Eq. (6.50). Next, the experiences are randomly sampled, typically in batches, to learn $Q^*(s, a)$ per Eqs. (6.46)–(6.49). This is known as **experience replay**, and is facilitated by DQN being an off-policy algorithm.

$$L(\theta) = \frac{1}{2} \left(Q^\pi_{\text{tar}}(s, a) - Q^{\pi_\theta}(s, a) \right)^2 \tag{6.48}$$

$$\theta = \theta - \eta \nabla_\theta L(\theta) \tag{6.49}$$

$$\mathcal{D} = e_1, e_2, ..., e_N \text{ where } e_t = (s_t, a_t, r_t, s_{t+1}) \tag{6.50}$$

One specific challenge related to this training method is the question of exploration vs. exploitation. If the state-space is insufficiently explored, many $Q^\pi(s, a)$ pairs remain inaccurate, and possibly no better than random initialization. On the other hand, in order for $Q^\pi(s, a)$ to progress toward $Q^*(s, a)$, the agent must exploit any information learned during previous training iterations. A common solution is the ε-*greedy* policy, in which the agent selects the greedy action with probability $1 - \varepsilon$, and a random action with probability ε. When beginning training iterations, ε is set high, e.g. close to 1.0; as training progresses, ε is reduced such that only the greedy actions are taken after sufficient exploration of the state-space.

An alternative to ε-greedy is the *Boltzmann policy*. It improves over the random exploration by selecting actions based upon their relative Q-values; actions with higher Q-values are explored, and those with lower Q-values are essentially ignored. These are chosen by constructing a probability distribution over the

Q-values for all actions in state s using the softmax function. This is described in Eq. (6.51), where the temperature parameter τ controls the spread.

$$P(a \mid s) = \frac{\frac{e^{Q^{\pi}(s,a)}}{\tau}}{\sum_{a'} \frac{e^{Q^{\pi}(s,a')}}{\tau}} \tag{6.51}$$

6.6.3.5 Advantage Actor-Critic

Thus far, we have discussed two methods to learn a policy which maximizes the objective, namely policy functions and value functions. It should be reassuring that we can in fact improve upon either method by combining them. We motivate this by looking at the policy gradient shown in Eq. (6.42), where the policy π_θ is reinforced by the return $R(\tau)$. Recall that $\nabla_\theta J(\pi_\theta)$ is learned using Monte Carlo sampling of single trajectories. This presents two potential challenges. First, the gradients exhibit high variance as the trajectories sampled may diverge at times. Second, the return may approach 0 over T, in which case the gradient approaches 0. In this scenario, *good* policies are not reinforced nor *bad* policies discouraged. To mitigate these, we instead reinforce π_θ using a learned function, which evaluates each action with respect to the current policy's average action for a given state. This is shown in Eq. (6.52), where the policy is reinforced by the *advantage* function shown in Eq. (6.53). Equation (6.52) is considered an **actor-critic** method, because we have two entities learning jointly. The **actor** learns the parameterized policy π_θ, and the **critic** learns the value function A^π used to evaluate state-action pairs. This is illustrated in Figure 6.20, where the agent entity in Figure 6.17 is now shown as a combination of actor and critic.

$$\nabla_\theta J(\pi_\theta) = \mathbb{E}_{\tau \sim \pi_\theta}\left[\sum_{t=0}^{T} A_t^\pi \nabla_\theta \log\left(\pi_\theta(a_t \mid s_t)\right)\right] \tag{6.52}$$

$$A^\pi = Q^\pi(s_t, a_t) - V^\pi(s_t) \tag{6.53}$$

$$\mathbb{E}_{a \in A}[A^\pi(s_t, a) = 0] \tag{6.54}$$

There are a few key concepts folded into Eq. (6.53). First, if all actions are equivalent, then $A^\pi = 0$, whereby all action probabilities remain unchanged; this is shown by Eq. (6.54). Second, A^π becomes a relative measure by comparing $Q^\pi(s_t, a_t)$ with $V^\pi(s_t)$. As such, specific actions are not penalized when π_θ is in a particularly bad state, or incorrectly encouraged when π_θ is in a particularly good state.

Naturally, the next question is how to estimate A^π? Per Eq. (6.53), this requires knowledge of both Q^π and V^π. Rather than learn them both, it is more efficient to learn V^π and use the rewards from n time steps of a single trajectory to estimate Q^π.

Figure 6.20 Actor-critic method.

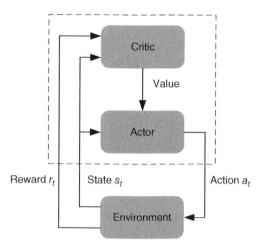

This is shown in Eq. (6.55), where $\widehat{V}^{\pi}$ is assumed learned by the critic. This results in the *n-step* estimate of A^{π} shown in Eq. (6.56). Clearly n is a user-defined parameter, which must be tuned for the desired variance and bias. Small n results in low-variance and high-bias, while large n results in the opposite.

$$Q^{\pi}(s_t, a_t) \approx r_t + \gamma r_{t+1} + \gamma^2 r_{t+2} + \ldots + \gamma^n r_{t+n} + \gamma^{n+1} \widehat{V}^{\pi}(s_{t+n+1})$$

$$(6.55)$$

$$A^{\pi}_{\text{NSTEP}}(s_t, a_t) \approx r_t + \gamma r_{t+1} + \gamma^2 r_{t+2} + \ldots + \gamma^n r_{t+n} + \gamma^{n+1} \widehat{V}^{\pi}(s_{t+n+1}) - \widehat{V}^{\pi}(s_t)$$

$$(6.56)$$

Similar to policy gradients and Q-learning, it is common practice for both the actor and critic to be modeled using an artificial neural network. This is illustrated in Figure 6.21 for the A2C method, assuming N hidden layers, and M possible actions per state. Similar to many modern architectures, we assume the representation of the Nth hidden layer is representative of both $\pi(a_t, s_t)$ and $V^{\pi}(s_t)$ with respect to the input state representation. As such, these outputs diverge in successive network layers only. Having already discussed the policy gradient shown in Eq. (6.52), the remaining question is how to learn V^{π}? The answer is exactly the same as $Q^{\pi}(s_t, a_t)$ as was discussed in Deep Q-Networks. In fact, $V^{\pi}_{\text{tar}}(s_t)$ is defined similarly to $Q^{\pi}_{\text{tar}}(s_t, a_t)$ and is shown in Eq. (6.57).

$$V^{\pi}_{\text{tar}}(s) = r + \widehat{V}^{\pi}(s')$$

$$(6.57)$$

6.6.3.6 Example Reinforcement Learning Applications

In [74], Chang et al. propose an RL approach to SU DSA for both the Advanced Wireless Services 3 (AWS-3) and 3.5 GHz ISM bands. Both were recently opened to SUs on a non-interfering basis with incumbent federal systems. This is

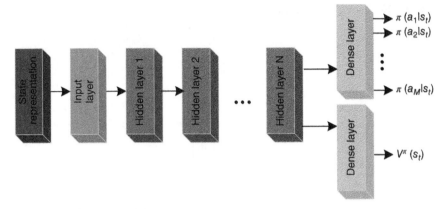

Figure 6.21 Deep advantage actor-critic network.

especially difficult as each SU is assumed to have no a priori knowledge regarding the behavior of the multiple PUs and other SUs. Their specific approach is to pair DQNs with reservoir computing (RC) to learn an effective DSA strategy. The size of the state-space is dictated the number of available channels and number of PUs. As this can become quite large, a DQN is used to approximate the Q-function. Also, to learn temporal dependencies, deal with imperfect sensing information, and to reduce training complexity, the DQN is modeled using an RNN based echo state network (ESN). While training online, when each SU accesses an idle channel, or collides with another SU, the reward is based upon the achievable data transmission rate. When the SU collides with a PU, the SU is penalized (negative reward). When the SU is idle, the reward is 0.

In [75], Tathe and Sharma propose an RL approach to resource scheduling in LTE-Advanced. Their motivation is to optimize system capacity, system throughput, and spectral efficiency, which depends heavily upon the resources allocated to each UE from a BS. Since resources are allocated on a subframe basis, evaluation of scheduling algorithms on this time scale carries significant complexity. This can be mitigated by modeling the selection process as an MDP. Their specific approach is to use an actor-critic method to learn an optimum selection policy from the pool of round robin, proportional Fair, best channel quality indicator, maximum rate, maximum-largest weighted delay first, and exponential/proportional fair. Although the actions are clearly the specific scheduling algorithms, the authors do not discuss the specific rewards used during training. It is however conceivable that these are specific to each scheduling algorithm, but also scaled to account for differences in numerical representation.

In [76], Wang et al. propose several multi-agent RL (MARL) algorithms to optimize channel allocations and maximize resource utilization within 5G wireless networks. They argue that in order for a 5G network to meet the anticipated throughput and QoS requirements under continued spectrum scarcity constraints, the network itself must be cognitive in nature. Each cell is therefore considered a multi-agent system (MAS) comprised of a base station (BS), and multiple PU and SU entities; each entity is an individual agent, and all others, that agent's environment.

To optimize channel allocations, each PU learns to reallocate its unused resources to SUs when available. It does so by collaborating with other PUs to determine aggregate unused resources, and with SUs to determine resource requirements. Accordingly, each SU is trained to obtain resources to meet its individual communications needs. Both entities model their decision-making using an MDP, where rewards are maximum for optimal resource allocation strategies.

To maximize resource utilization, each BS learns to load balance with other BSs in its immediate area. That is, when the distribution of channel resources exceeds capacity, a BS may switch to an underutilized channel, and offer its previous channel resources to another BS. It also models its decision-making using an MDP, where rewards are maximum when resource utilization is also maximum.

6.7 Looking Forward

Further advances in automation, machine learning, and artificial intelligence should only increase our expectations regarding intelligent radio capabilities. Many of the use cases discussed earlier have been deployed or are clearly within reach today. Maybe the appropriate question to ask is *"what additional services can be provided through wireless infrastructure?"* Global connectivity may indeed be an enabler for many applications providing social and economic benefits, which brings us back to the major theme of this text: wireless coexistence. How efficiently can we utilize the spectrum? In attempting to answer this question, we explored several concepts, architectures, and algorithms to improve both spectrum utilization and user experience. Furthermore, the CRM framework has the ability to quickly harness new learning techniques within the toolbox that the core can utilize to improve optimization and decision-making. This, coupled with the evolution of wireless technology toward larger networks, and smaller, more capable devices, serves only to fuel our imagination.

References

1 DySPAN-SC. (2019). IEEE Standard 1900.1-2019, IEEE, New York.

2 DySPAN-SC. (2009). IEEE Standard 1900.4-2009, IEEE, New York.

3 DySPAN-SC. (2011). IEEE Standard 1900.4a-2011, IEEE, New York.

4 TensorFlow. (2020). https://www.tensorflow.org (accessed 9 June 2020).

5 B. A. I. Research. (2020). Caffe, Berkely Artificial Intelligence Research. https://caffe.berkeleyvision.org (accessed 9 June 2020).

6 PyTorch. (2020). https://pytorch.org. (accessed 9 June 2020).

7 Microsoft. (2020). Microsoft Cognitive Toolkit. https://docs.microsoft.com/en-us/cognitive-toolkit (accessed 9 June 2020).

8 Scikit-Learn. (2020). https://scikit-learn.org/stable (accessed 9 June 2020).

9 Mitola, J. (2000). *Cognitive Radio: An Integrated Agent Architecture for Software Defined Radio*. Stockholm: KTH Royal Institute of Technology.

10 Hayken, S. (2005). Cognitive radio: brain-empowered wireless communications. *IEEE Journal on Selected Areas in Communications* 23 (2): 201–220.

11 Seetoodeh, P. and Haykin, S. (2017). *Fundamentals of Cognitive Radio*. Hoboken: Wiley.

12 FCC. (2002). Spectrum Policy Task Force ET Docket No. 02-135, FCC, Washington D.C.

13 Wyglinski, A.M., Nikovee, M., and Hou, T.Y. (2010). *Congnitive Radio Communications and Networks*. Burlingon: Academic Press.

14 Marshall, P. (2010). *Quantitative Analysis of Cognitive Radio and Network Performance*. Norwood: Artech House.

15 Buljore, S., Harada, H., Filin, S. et al. (2009). Architecture and enablers for optimized radio resource usage in heterogeneous wireless access networks: the IEEE 1900.4 working group. *IEEE Communications Magazine* 47: 122–129.

16 3GPP. (2015). TR 36.889 V13.0.0, 3GPP, Valbonne.

17 Ericsson Mobile. (2019). Ericsson Mobility Report 2019, Ericsson, Stockholm.

18 Bhandari, S. and Joshi, S. (2018). Cognitive radio technology in 5G wireless communications. *2nd IEEE International Conference on Power Electronics, Intelligent Control and Energy Systems*, New Delhi.

19 Amazon. (2020). Amazon Alexa. https://developer.amazon.com/en-US/alexa (accessed 9 June 2020).

20 Microsoft. (2020). Microsoft Cortana. https://www.microsoft.com/en-us/cortana (accessed 9 June 2020).

21 Google. (2020). Google Assistant. https://assistant.google.com (accessed 9 June 2020).

22 Apple. (2020). Apple Siri. https://www.apple.com/siri (accessed 9 June 2020).

23 Cisco. (2020). Webex. https://www.webex.com (accessed 9 June 2020).

24 Zoom Technologies Inc. (2020). Zoom. https://zoom.us (accessed 9 June 2020).

25 Netflix, Inc. (2020). Netflix. https://www.netflix.com (accessed 9 June 2020).

26 YouTube. (2020). YouTube. https://www.youtube.com (accessed 9 June 2020).

27 Google Inc. (2020). Google Maps. https://www.google.com/maps (accessed 9 June 2020).

28 Waze Mobile. (2020). Waze. https://www.waze.com (accessed 9 June 2020).

29 MapQuest. (2020). Mapquest. https://www.mapquest.com (accessed 9 June 2020).

30 Lustosa, B. (2018). Apple's iOS update frequency has increased 51% under Cook's management, Venture Beat. https://venturebeat.com/2018/02/28/apples-ios-update-frequency-has-increased-51-under-cooks-management (accessed 9 June 2020).

31 Owen, M. (2020). Analyst estimates average lifespan for all Apple devices at over four years, ZDNet. https://appleinsider.com/articles/18/03/01/analyst-estimates-average-lifespan-for-all-apple-devices-at-over-four-years (accessed 9 June 2020).

32 FCC. (2020). Cell Phone Fraud. https://www.fcc.gov/sites/default/files/cell_phone_fraud.pdf (accessed 24 March 2020).

33 Sowa, J. F. (2015). The cogntive cycle. *Proceedings of the Federated Conference on Computer Science and Information Systems*, Lodz.

34 Campbell, P.L. (2011). *Pierce, Pragmatism, and the Right Way of Thinking*. Albuquerque: Sandia National Laboratories.

35 Pietarinen, A.-V. and Bellucci, F. (2015). New light on Pierce's conceptions of Retroduction, deduction, and scientific reasoning. *International Studies in the Philosophy of Science* 28 (4): 353–373.

36 Stanford Encyclopedia of Philosophy. (2020). Charles sanders pierce, Center for the Study of Language and Information, Stanford University, 2020. https://plato.stanford.edu/entries/peirce (accessed 27 March 2020).

37 Brown, I.T. (2018). *A New Conception of War: John Boyd, the U.S. Marines, and Maneuver Warfare*. Quantico: Marine Corp University Press.

38 Farahi, F. and Yazdi, H.S. (2020). Probabilistic Kalman filter for moving object tracking. *Signal Processing: Image Communciations* 82: 115751.

39 Kulkarni, A. and Rani, E. (2018). Kalman filter based multiple object tracking system. *International Journal of Electronics, Communication & Instrumentation Engineering Research and Development* 8 (2): 1–6.

40 Steketee, M., Miyoaka, A., and Spiegelman, M. (2015). *Social Network Analysis*. Jenison: Elsiever.

41 Abbas, N., Nasser, Y., and El Ahmad, K. (2015). Recent advances on artificial intelligence and learning techniques in cognitive radio. *EURASIP Jorunal on Wireless Communications and Networking* 2015: 174.

42 Quora. (2019). What Exactly Is Machine Intelligence?, Forbes. https://www.forbes.com/sites/quora/2019/11/15/what-exactly-is-machine-intelligence/#2111e88d187c (accessed 10 June 2020).

43 Hankiewitz, K. (2018). 10 examples of automation, Medium. https://medium.com/@kamila/10-examples-of-automation-7326ef303772 (accessed 10 June 2020).

44 Fitzek, F.H. and Katz, M.D. (2007). *Cognitive Wireless Networks*. Dordrecht: Springer.

45 Cai, T., Kourdouridis, G. P., Johansson, J. et al. (2010). An implementation of cognitive resource manager on LTE platform. *21st Annual IEEE Symposium on Personal, Indoor, and Mobile Radio Communications*, Istanbul.

46 Patra, A., Achtzehn, A., and Mahonen, P. (2015). ULLA-X: a programmatic middleware for generic cognitive radio network control. *2015 IEEE International Symposium on Dynamic Spectrum Access Networks (DySPAN)*, Stockholm,.

47 Gavrilovska, L., Atanasovski, V., Macaluso, I., and DaSilva, L.A. (2013). Learning and reasoning in Cogntive radio networks. *IEEE Communication Surveys and Tutorials* 15 (4): 1761–1777.

48 Alpaydin, E. (2014). *Introduction to Machine Learning*. Cambridge: MIT Press.

49 Goodfellow, I., Bengio, Y., and Courville, A. (2016). *Deep Learning*. Cambridge: MIT Press.

50 Bishop, C.M. (2006). *Pattern Recognition and Machine Learning*. Cambridge: Springer.

51 Mazur, M. (2015). A step by step backpropagation example. https://mattmazur. com/2015/03/17/a-step-by-step-backpropagation-example (accessed 3 June 2020).

52 Edwards, C.H. and Penney, E.E. (1986). *Calculus and Analytic Geometry*. Englewood Cliffs: Prentice Hall.

53 He, K., Zhang, X., Ren, S. et al. (2016). Deep residual learning for image recognition. *IEEE Conference on Computer Vision and Pattern Recognition*, Las Vegas.

54 Bengio, Y., Simard, P., and Frasconi, P. (1994). Leaming long-term dependencies with gradient descent is difficult. *IEEE Transactions on Neural Networks* 5 (2): 157–166.

55 Hochreiter, S. and SchmidHuber, J. (1997). Long short term memory. *Neural Computation* 9 (8): 1735–1780.

56 Colah, C. (2015). Understanding LSTM networks, https://colah.github.io/posts/ 2015-08-Understanding-LSTMs (accessed 4 June 2020).

57 Mishra, D. (2020). Transposed convolution demystified. Towards Data Science. https://towardsdatascience.com/transposed-convolution-demystified-84ca81b4baba (accessed 13 June 2020).

58 Kingma, D. P. and Welling, M. (2014). Auto-encoding variational bayes. https:// arxiv.org/pdf/1312.6114.pdf (accessed 17 June 2020).

59 Doerch, C. (2016). Tutorial on variational autoencoders. https://arxiv.org/abs/ 1606.05908 (accessed 9 June 2020).

60 Makhzani, A., Goodfellow, I., Shlens, J. et al. (2016). Adversarial autoencoders. https://arxiv.org/pdf/1511.05644v2.pdf (accessed 17 June 2020).

61 Goodfellow, I. J., Pouget-Abadie, J., Mirza, M. et al. (2014). Generative adversarial nets. https://arxiv.org/pdf/1406.2661v1.pdf (accessed 17 June 2020).

62 Chew, D. and Cooper, B. (2020). Spectrum sensing in interference and noise using deep learning. *54th Annual Conference on Information Sciences and Systems (CISS)*, Princeton.

63 Liu, X., Yang, D. and El Gamal, A. (2017). Deep neural network architectures for modulation classification. *51st Asilomar Conference on Signals, Systems, and Computers*, Pacific Grove.

64 O'Shea, T., Karra, K. and Clancy, C. T. (2017). Learning approximate neural estimators for wireless channel state. *IEEE International Workshop on Machine Learning for Signal Processing*, Tokyo.

65 Rajendran, S., Meerty, W., Lendersz, V. et al. (2018). SAIFE: unsupervised wireless spectrum anomaly detection with interpretable features. *International Symposium on Dynamic Spectrum Access Networks*, Seoul.

66 Akbar, I.A. and Tranter, W.H. (2007). Dynamic spectrum allocation in cognitive radio using hidden markov models: poisson distribution case. *IEEE SoutheastCon*, Richmond.

67 Senthilkumar, S. and Geetha, P.C. (2018). Hidden Markov model based channel selection framework for cognitive radio network. *Computers and Electrical Engineering* 65: 516–526.

68 Li, X. and Xiong, C. (2014). Markov model band for heterogenous cognitive radio networks with multiple dissimilar users and channels. *International Conference on Computing, Networking and Communications*, Honolulu.

69 Graesser, L. and Keng, W.L. (2019). *Foundations of Deep Reinforcement Learning.* Boston: Addison-Wesley.

70 Russell, S.J. and Norvig, P. (2010). *Artificial Intelligence: A Modern Approach.* Upper Saddle River: Prentice Hall.

71 Kapoor, S. (2018). Policy gradients in a nutshell. https://towardsdatascience.com/policy-gradients-in-a-nutshell-8b72f9743c5d (accessed 5 July 2020).

72 Sutton, R.S. and Barto, A.G. (2018). *Reinforcement Learning: An Introduction.* Cambridge: MIT Press.

73 Mnih, V., Kavukcuoglu, K., Silver, D. et al. 2013. Playing atari with deep reinforcement learning. https://arxiv.org/abs/1312.5602 (accessed 28 June 2020).

74 Chang, H.-H., Song, H., Yi, Y. et al. (2019). Distributive dynamic Spectrum access through deep reinforcement learning: a Resevoir computing based approach. *IEEE Internet of Things Journal* 6: 1938–1948.

75 Tathe, P.K. and Sharma, M. (2018). Dynamic actor-critic: reinforcement learning based radio resource scheduling for LTE-advanced. *2018 Fourth International Conference on Computing Communication Control and Automation*, Pune, India, pp. 1–4. doi: https://doi.org/10.1109/ICCUBEA.2018.8697808.

76 Wang, D., Song, B., Chen, D., and Du, X. (2019). Intelligent cognitive radio in 5G: AI-based hierarchical cognitive cellular networks. *IEEE Wireless Communications* 26: 54–61.

7

Coexistence Standards in IEEE 1900

7.1 DySPAN Standards Committee (IEEE P1900)

The P1900 group was first established in 2005 as a joint working group between the IEEE Communications Society (COMSOC) and the IEEE Electromagnetic Compatibility Society (EMC) with the goal of developing standards specifically targeted for advanced radios capable of performing spectrum management. With the booming success of third-generation cellular technologies in the rear view and fourth-generation standards already in development, it was clear that there was going to be no conceivable end to the world's appetite for mobile data services and, by extension, wireless spectrum resources. By this time, it was clear that the leasehold model employed by spectrum regulatory agencies in the past was providing an extremely inefficient use for their nation's limited spectrum resources. Lease holders would often deploy in a limited geographic area, leaving large swaths of the scarce spatiofrequency spectrum unused.

In creating this new body, the goals of the COMSOC and the EMC were to foster new technologies that could employ advanced techniques for using the idle spectrum as efficiently as possible. There were two primary goals behind the P1900 Standards Committee. First, the committee sought to develop standards that would allow new radios the ability to interact intelligently and dynamically to maximize the spectral resource usage in a given geographic area using a combination of environmental sensing, pre-established policy frameworks, and cognitive reasoning. The second, and arguably more important goal, was to avoid interference with the primary users of any allocated spectrum. The majority of these legacy spectrum users gained their leasehold rights through government auctions and were guaranteed by the terms of those auctions that the spectrum resources would be available to them at all times. This means that efficient allocation of the spectrum will, by necessity, rely on intelligent, policy-based methods of coordination.

Wireless Coexistence: Standards, Challenges, and Intelligent Solutions, First Edition.
Daniel Chew, Andrew L. Adams, and Jason Uher.
© 2021 The Institute of Electrical and Electronics Engineers, Inc.
Published 2021 by John Wiley & Sons, Inc.

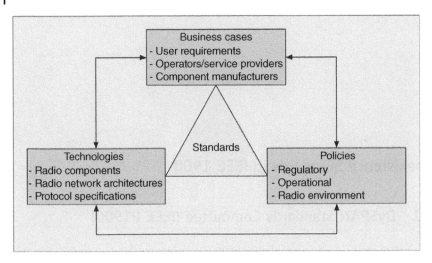

Figure 7.1 Interactive standards model for 1900 [2]. *Source*: Buljore et al. [3]. (© [2020] IEEE).

At its core, the P1900 standards are intended to provide a guide for reasoning about how wireless coexistence can function in a practical way [1]. The original intention of the P1900 standards was not to simply provide a "wireless coexistence" standard that would allow sharing of the spectrum but to provide an "interactive process of business case, technology, and policy" as the model for development as shown in Figure 7.1.

At the time, it was recognized that this standards committee would be required to look at the bigger picture when it came to the standards they were to generate [2]. To be successful, the committee would have to examine both ongoing business trends and regulatory policy developments to provide a useful roadmap for ensuring that *dynamic spectrum access* (DSA) would be possible. The committee was therefore focused on creating standards that emphasized architectures to facilitate network management and spectrum sharing between different technologies rather than specific *media access control* (MAC) and *physical* layer (PHY) functionality [4]. In addition, the committee would have to work closely with regulatory agencies, other standards bodies, and the commercial and academic research communities to ensure relevance and provide the best chance of ensuring interoperability between different systems. Figure 7.2 shows these relationships with respect to the P1900 standards committee. This map shows that the 1900 standards are informed by academic research while maintaining active relationships with the commercial and regulatory bodies.

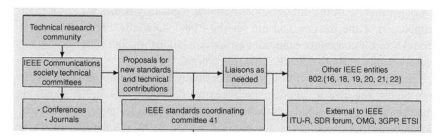

Figure 7.2 IEEE 1900 relationship to other technical contributors. *Source*: Granelli et al. [2]. (© [2020] IEEE).

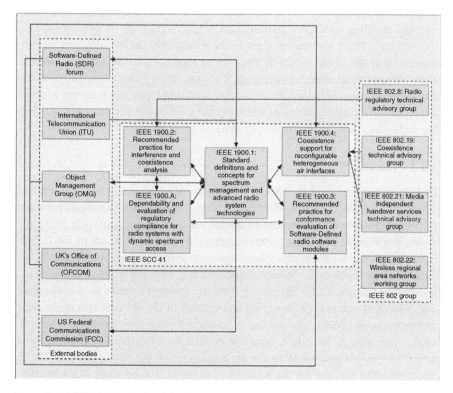

Figure 7.3 IEEE 1900 relationship with standards bodies and regulatory agencies. *Source*: Prasad et al. [5]. (© [2020] IEEE).

Cooperation with the different international regulatory bodies is critical to the completion of the mission laid out for the P1900 standards. Figure 7.3 demonstrates how the committee maintains an active relationship both with relevant regulatory bodies and other IEEE standards groups.

These active relationships ensure compatibility between international regulatory policy, the abstract models, architectures defined in the P1900 standards, and the various IEEE standards focused on specific access technologies. This relationship highlights one of the key motivations the P1900 standards, which is to provide an abstract model for wireless coexistence that can be used for regulatory and legal discourse without tying decisions to a specific technology.

The P1900 standards focus on three main methodologies to improve spectrum usage: fostering development of novel methods for *dynamic spectrum access* (DSA), defining frameworks for new radio systems with a focus on spectrum sharing, and enabling intelligent coordination of spectrum usage in existing wireless technologies. These goals are evident in the structure of the working groups, outlined later in this chapter, which when taken together provide a broad, cohesive vision of spectrum reuse technologies. The standards provided by the working groups provide a unified plan for improving spectrum usage and take into account the adoption of new technologies while working within the limitations of both legacy technology and legacy regulatory systems.

7.1.1 History

Following their creation in 2005, the P1900 working groups released the initial 1900.1 [6] and 1900.2 [7] specifications in 2008. These two specifications provided the initial groundwork for building dynamic radio systems capable of meeting the committee's goals. The 1900.1-2008 standard, titled "IEEE Standard for Definitions and Concepts for Dynamic Spectrum Access: Terminology Relating to Emerging Wireless Networks, System Functionality, and Spectrum Management" provided a framework for discussing dynamic spectrum radios in an organized and consistent manner. By providing an overview of the key concepts in *dynamic spectrum access* (DSA) systems and a consistent taxonomy for those concepts, 1900.1 was able to provide a common ground for comparing and contrasting new DSA system concepts. With similar goals in mind, the 1900.2 standard "IEEE Recommended Practice for the Analysis of In-Band and Adjacent Band Interference and Coexistence Between Radio Systems" defines a method for consistently comparing the amount of interference radio systems' exert on one another.

As the initial 1900.1 and 1900.2 standards were being released, the P1900 Standards Committee was reorganized into the Standards Coordinating Committee 41 (SCC41), also known as the dynamic spectrum access networks committee, or DySPAN. This new committee was still managed by the COMSOC and EMC groups, and carried the IEEE 1900 numbering scheme. Following this transition, initial versions of the 1900.4 [8], 1900.5 [9], and 1900.6 [10] standards were released. These releases defined a model for intelligent-coordinated spectrum usage between networks, a formalized policy-based control mechanism, and the communication interface used when sharing data between spectrum sensors and DSA networks.

In 2011, the IEEE 1900 standards was brought back under the purview of the COMSOC and adopted into the IEEE Communications Society Standards Board (CSSB) as the IEEE DySPAN Standards Committee (DySPAN-SC). Since then, the working groups have produced amendments to the different standards to allow for changing paradigms in DSA networks. These amendments include allowances for the rapid advance in Artificial Intelligence (AI) and Machine Learning (ML) as well as the availability of new *TV White space* spectrum. This new spectrum, often devoted to DSA radios, was freed up in several countries around the world as older analog terrestrial broadcast systems were decommissioned in favor of more efficient digital systems. In addition to these amendments, the 1900.7 working group released its initial standard in 2015 [11]. This working group provides standard *medium access control* (MAC) and *physical* (PHY) layer specifications allowing radios to operate in white spaces while ensuring a minimal impact to incumbent users should they need to utilize the spectrum.

Work continues on the IEEE 1900 standards with the latest version of IEEE 1900.1 released in 2019 (IEEE 1900.1-2019) [12]. The standard IEEE 1900.4-2009 was withdrawn on 5 March 2020. The amendments 1900.4a-2011 and 1900.4.1-2013 are still active.

7.1.2   The Working Groups' Overview

This section provides an overview of the six active working groups within the IEEE 1900 DySPAN-SC committee. While there were originally seven working groups, the P1900.3 group was disbanded, and its mission is no longer actively pursued by the DySPAN-SC. The 1900.3 group was tasked with providing a methodology for estimating the conformance of software-defined radio (SDR) systems with respect to a given published standard. The goal was to enable SDR systems to achieve the same level of technical certification currently required for hardware radio systems. In the end, this goal was likely beyond the ability of software validation methods of the time and the group was disbanded. Figure 7.4 shows the hierarchy of the different 1900 standards and the working groups that run them.

7.1.3   1900.1 Working Group

The 1900.1 working group is focused on "Definitions and Concepts for Dynamic Spectrum Access: Terminology Relating to Emerging Wireless Networks, System Functionality, and Spectrum Management." The field of potential problems in DSA radio technologies is both broad and nuanced, which necessitates complex

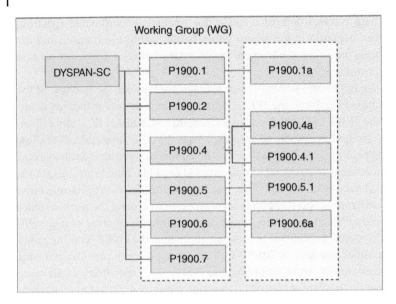

Figure 7.4 IEEE 1900 working groups and published standards. *Source*: Harada et al. [1]. (© [2020] IEEE).

solutions that can often be difficult to evaluate and compare objectively. The primary goal of 1900.1 is to provide a common language and basic topological model for technologies related to DSA radios. By providing both a baseline model and taxonomy for DSA systems, analysts are able to establish a grounded methodology for comparing the pros and cons of candidate technologies in a systematic and repeatable fashion.

What the 1900.1 standard provides is a way to talk about and analyze the separate functional domains of a DSA systems in a plain and consistent manner. The standard provides clear definitions for different radio classes, radio traits, spectrum access methods, user roles, and more. The models and definitions are sufficiently decomposed that it should be possible to map any proposed DSA scheme into the common taxonomy. With this common language, it should be clear what a particular solution's capabilities and limitations are, and how it fits into an overall DSA radio system. Figure 7.5 from [6] shows an example of this functional decomposition as presented by the 1900.1 working group which outlines the different functions and actions that a DSA radio or network would use when communicating.

This model also helps a designer determine how best to assemble a DSA-based system. The figure below, taken from the 1900.1 specification, shows a high-level overview of the concepts the working group seeks to codify with the models and ontology.

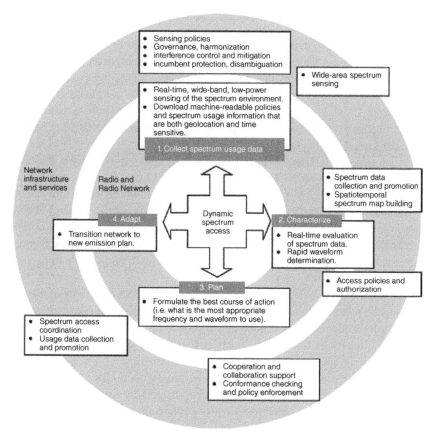

Figure 7.5 Functional decomposition of a DSA radio from 1900.1. *Source*: IEEE Std 1900.1-2008 [6]

In each step of the DSA loop, the radio system must collect a piece of information or take some action to accomplish the high-level goals. In any given DSA system, it is unlikely that each of those pieces of information will come from the same subsystem or that those actions will be taken by the same piece of equipment. By applying the ontology and concepts from the 1900.1 standard, a DSA system designer can be assured that each of the required functions is accounted for by mapping the functions of the selected components to the 1900.1 model.

The 1900.1 specification, 1900.1-2008, was originally released in 2008 and amended with 1900.1a-2012. The most recent version of the 1900.1 specification is 1900.1-2019. This includes all updates from the 1900.1 working group and is the primary foundational reference for the 1900 series.

7.1.4 1900.2 Working Group

The 1900.2 working group is focused on the "Recommended Practice for the Analysis of In-Band and Adjacent Band Interference and Coexistence Between Radio Systems." This standard provides an in-depth methodology for evaluating the impact of radio systems operating simultaneously, both for in-band operation and adjacent band operation. This standard, like the common taxonomy of 1900.1, provides a standard way of comparing the benefits and drawbacks of potential systems in a rigorous fashion.

The methodology described in the standard is unique in the sense that it does not solely examine the physical-layer interference caused by the radio systems. The approach forces analysts to take a holistic view of the radio networks and assess the overall impact to the network rather than simply the presence of energy above a threshold. Figure 7.6 from [7] demonstrates the process used to assess the overall impact of an interference event. Rather than simply collecting a count of events, the events are analyzed to determine if they actually caused harm to the impacted system at the application level.

One of the primary concerns with the adoption of the DSA technologies is that they have the potential to interfere with the incumbent users of a given spectrum allocation. By applying the methodology outlined in this standard, an analyst can evaluate the impact on the quality of service that a DSA user operating as a secondary user could incur on the incumbent radio network.

The methodology in 1900.2 can be used to analyze the coexistence capabilities of two DSA-based radio networks in the same band. This use case, opposed to the incumbent/secondary use case, considers that there may be a net benefit that arises from coexistence when there is not a clear right to spectrum availability. One example of this is in the case of the Wi-Fi (802.11) and Bluetooth (802.15.1). Both technologies operate in the same ISM bands and would normally cause harmful interference to one another. By employing a coexistence protocol, such as the 802.15.2 standard, the two are able to successfully operate at the same time. The 1900.2 methodology is equipped to analyze the impacts, both positive and negative, to the performance of these two technologies.

The primary focus of this standard is on ensuring that the methodology is properly applied and documented which allows a fair comparison between proposed DSA technologies. The process involves four stages: defining the scenario, establishing the criteria for interference, establishing the higher layer models, and performing analysis and simulations. For each stage of this process, the standard outlines best practices and common pitfalls that an analyst should keep in mind when assessing the coexistence capabilities of a new DSA technology.

The original release of the 1900.2 standard was published in 1998 and is still listed as active on the IEEE standards website.

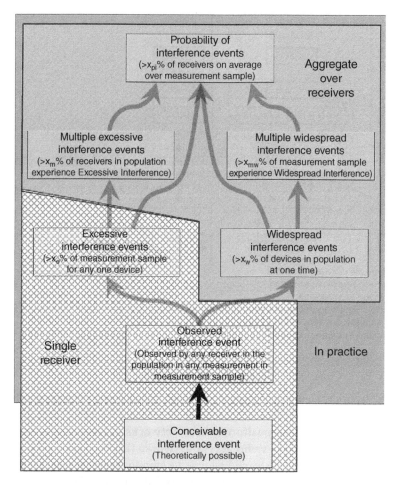

Figure 7.6 Interference assessment flowchart. *Source*: I. 1. S. Committee [7].

7.1.5 1900.4 Working Group

The IEEE 1900.4 "Working Group on Architectural Building Blocks Enabling Network-Device Distributed Decision Making for Optimized Radio Resource Usage in Heterogeneous Wireless Access Networks" attempts to define a model for enabling distributed decision making between heterogeneous networks to improve the overall capacity of the networks. The protocol outlined in this standard improves the overall *quality of service* (QoS) and spectrum usage of cooperating networks by ensuring that differing *radio access technologies* (RATs) are using their allocated resources in an optimal way [8].

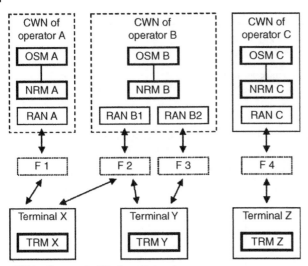

CWN – Composite Wireless Network
OSM – Operator Spectrum Manager
RAN – Radio Access Network
NRM – Network Reconfiguration Manager
TRM – Terminal Reconfiguration Manager
F 1, 2, 3, and 4 – Frequency Bands 1, 2, 3, and 4

Figure 7.7 Spectrum reconfiguration example. *Source*: I. 1. S. Committee [8].

Figure 7.7 from the standards document [8] is an example of an environment where 1900.4 can provide optimization abilities. In this environment, there are three operators managing **composite wireless networks** (CWNs) which may contain a heterogeneous mix of **radio access networks** (RANs). Under 1900.4, each of these network operators would be able to monitor their own spectrum usage with their **operator spectrum manager** (OSM) utilizing data reported from the **terminal reconfiguration manager** (TRM) running on terminals within their networks. Based on this data, the **network reconfiguration managers** (NRM) can coordinate between the different operators to optimally allocate the spectrum resources they have available to the loads present on their RANs. Once the decision on how to allocate the resources has been made, the networks and terminals can be commanded to make the necessary adjustments through the NRMs and TRMs.

While this figure gives a high-level overview of the concepts and terminology used in 1900.4, it is not the only possible model for optimization. There are several use cases dictated in the standard that allow networks to optimize their resource usage including scenarios where the networks or terminals cannot be reconfigured (i.e. there is no NRM or TRM available).

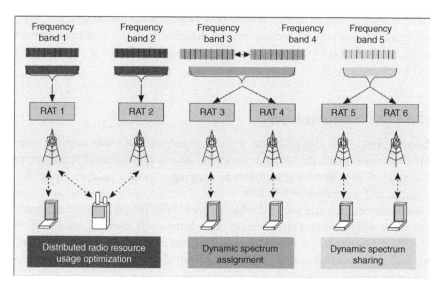

Figure 7.8 Dynamic spectrum management example [13]. *Source*: Harada et al. [1]. (© [2020] IEEE).

The standard provides three primary methods for facilitating DSA between heterogeneous network technologies: dynamic spectrum management, dynamic spectrum assignment, and distributed radio resource usage as shown in Figure 7.8 [13].

The first method, distributed radio resource usage optimization, puts the onus for flexibility solely in the hands of the user terminals. In this case, the NRMs generate policies for the terminals according to the available, but fixed, RAN resources and the TRMs utilize these policies to select radio resources for transmit. The second method, dynamic spectrum sharing, utilizes fixed frequency allocations for the individual RANs but allows the NRMs and TRMs to generate dynamic resource allocations according to the spectrum use policies in effect. The final method, dynamic spectrum assignment, feeds the band assignments "top down" from the OSM to the RANs. The OSM looks at the high-level spectrum assignments and policies to reallocate the RANs in the most optimal way.

The original standard, released in 2009, provided architectural and functional definitions of new elements that would need to be added to existing **radio access technologies** (RATs). These new components allow the RATs to coordinate resources by sharing information both vertically (within a single operator) and horizontally (between operators) at both the network-to-network level and at the radio-to-radio level. An amendment was released in 2011, 1900.4a, which adds specific support for operation of 1900.4 components within the TV white space bands [14].

Another project, released in 2013, is the 1900.4.1 specification that adds service interface definitions and protocol message formats designed to provide a common schema for exchanging 1900.4 related information between heterogeneous networks and RAT technologies [15].

7.1.6 1900.5 Working Group

The IEEE 1900.5 working group on "Policy Language and Architectures for Managing Cognitive Radio for Dynamic Spectrum Access Applications" facilitates an ecosystem of interoperable DSA radios by defining a "vendor-independent policy language and associated architecture."

Given the complex and unpredictable nature of DSA radios, it can be extremely difficult to design a robust specification that addresses all operating scenarios. As an alternative to this, the 1900.5 group focuses on defining a "Policy Language" that will allow intelligent radios to reason about what is (and is not) allowed in a given scenario.

As shown in Figure 7.9 [9] below, the goal of the policy language put forth in 1900.5 is to streamline policy compliance for everyone: the regulators, the network operators, users of the DSA radios, and the radio manufacturers. In this way, regulators can continue to specify the rules and stipulations on spectrum use in plain language, which is easily converted into a 1900.5 policy language easily ingested in a machine-readable format by network operators and DSA radios.

In addition to specifying the requirements and methods of a compliant policy language, the standard also gives a standard model for a radio system that would implement a **policy-based radio system** (PBRS). Figure 7.10 [1] shows a model for a PBRS that takes action inferred from policies, rather than by following a strict

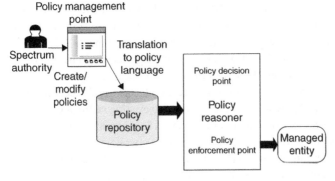

Figure 7.9 Policy-based control architecture. *Source*: I. 1. S. Committee [9].

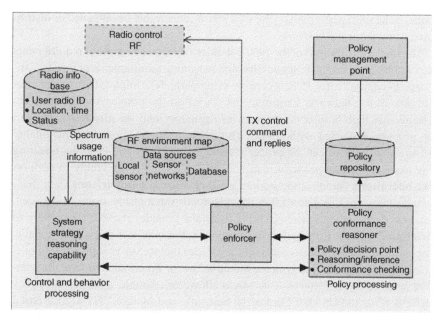

Figure 7.10 System model for a policy-based radio system (PBRS). *Source*: Harada et al. [1]. (© [2020] IEEE).

protocol. In this diagram, the machine-readable policy definitions are provided by a **policy management point** (PMP) and optionally stored in a **policy repository**. From here, the policies can be read by the **policy conformance reasoner** (PCR) who is responsible for evaluating requests from the radio system about what is or is not allowed given the current regulatory domain. Adherence to these policies, and ultimate configuration of the radio, is managed by the **policy enforcer** (PE) who receives reconfiguration requests from the **system strategy reasoning capability** (SSRC) in an attempt to optimize the current spectrum usage. The SSRC has the job of determining the currently optimal radio parameters given the current state of the network. This state consists of any information that the SSRC has about the network including the RF environment, network topology, transmitter and receiver characteristics, regulatory location, and so on. The SSRC can even query the PCR to determine the range of current constraints as an input into its decision-making process. It is important to note that these functional decompositions are designed such that the most computationally expensive portions of a policy-based system can be offloaded to centralized locations if necessary. Of the components outlined in the model, it is likely that the only component which must be physically colocated with the radio node is the PE. Actions which take place on a long time-scale relative to network maneuver such as reasoning actions taken by the SSRC or policy validation performed by the PCR could conceivably be offloaded or distributed throughout the network.

The final contributions of the 1900.5 standard include a number of requirements placed on the ontology, functionality, and reasoning requirements of a policy language to ensure that a PCR is capable of properly determining the current constraints on the network. Requirements placed on the ontology are designed to ensure that high-level concepts such as regulatory and security restriction can be properly expressed without ambiguity. The restrictions on the functionality ensure that policies can be properly reasoned about by the PCRs by ensuring the elements can be properly reasoned on. For example, the ability to perform **set operations** (union, intersection, etc.), **relational comparisons** (less than, greater than, etc.), and **spatial comparisons** (inside/outside, distance between, etc.) on data elements when making decisions provides the PCR the ability to determine not just if an action is permissible, but why it is (or is not). Finally, the standard places requirements on the policies themselves as well. For example, the standard dictates that policies should allow for inheritance to ease the complexity of policy definitions. This would allow, for example, a theoretical policy definition for the US FCC "Industrial Scientific and Medical" (ISM) unlicensed bands to inherit from a more general definition of unlicensed bands. These restrictions ensure that the PCR is able to reason about the relative importance of constraints when attempting to apply multiple policies simultaneously.

The first version of 1900.5, published in 2011, outlines the policy-based control architectures and language requirements to implement a policy-based radio system (PBRS) [9]. In 2017, an extension was released, 1900.5.2, which defines a vendor-independent approach for analyzing RF spectrum use [16]. This generalized method allows an analyst to determine the coexistence compatibility of radio networks composed of traditional radios, policy-based radio systems, or a combination of both.

7.1.7 1900.6 Working Group

The 1900.6 working group is on "Spectrum Sensing Interfaces and Data Structures for Dynamic Spectrum Access and other Advanced Radio Communication Systems." This standard seeks to define how spectrum sensors and the consumers of that data exchange information in a predictable and compatible fashion. To ensure compatibility across different network types and configurations, the working group takes great care to ensure that the interfaces for exchanging sensor data are independent of sensing methodology, the client radios, or the method by which the data consumers access the sensors.

The primary goal of the standard is to ensure that DSA radios can find the spectrum resources they need to communicate.

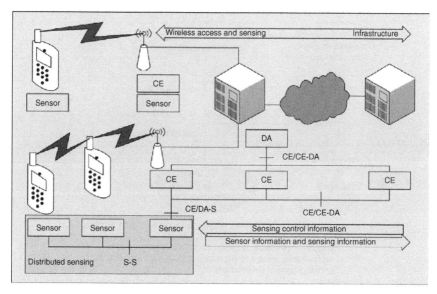

Figure 7.11 Cognitive network with multiple sensing sources. *Source*: Murroni et al. [18]. (© [2020] IEEE).

While the focus of cognitive radio research was once focused on the individual radios determining holes directly, recent advances have shown that utilizing a distributed model for locating spectral holes can provide a much more accurate result that also ensures compliance with regulatory specifications [17]. Figure 7.11 from [18] shows an example of a cognitive network with multiple *cognitive engines* (CE) that can utilize a number of different paradigms to obtain available spectral data. These include sensor data from the network itself, live measurements from nearby networks, and remote databases containing historical sensing data.

The 1900.6 standard defines the necessary communications protocols for exchanging data between two primary entity types: **producers of spectrum data** and **consumers of spectrum data**. Within the standard documentation, producers of data are always referred to as "Sensors." However, the consumers of data are typically referred to as either a **cognitive engine** (CE) or a **data archive** (DA). In the IEEE 1900 ontology, the CE is the portion of a DSA system that decides the transmission parameters to be used when participating in a network. The DA, on the other hand, is a repository of both historical sensor data and external information, such as regulatory parameters and policy information. Though the standard regularly refers to these as separate entities, the actual interface is defined the same.

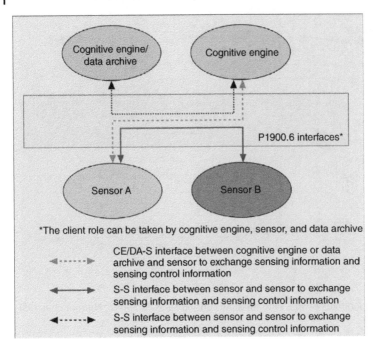

*The client role can be taken by cognitive engine, sensor, and data archive

CE/DA-S interface between cognitive engine or data archive and sensor to exchange sensing information and sensing control information

S-S interface between sensor and sensor to exchange sensing information and sensing control information

S-S interface between sensor and sensor to exchange sensing information and sensing control information

Figure 7.12 1900.6 participants and communication interfaces. *Source*: Harada et al. [1]. (© [2020] IEEE).

Figure 7.12 [1] shows the relationship between the three potential participant types, with the black dashed lines representing the three communication interfaces defined in the specification: *cognitive engine-to-sensor* (CE-S), *cognitive engine-to-cognitive engine* (CE-CE), and *sensor-to-sensor* (S-S) links. The CE-S link defined in this standard will likely be the primary method by which the radios obtain spectrum sensor data. CE-CE link is used to share data between network devices. If, for example, one CE is not able to communicate with a spectrum sensor, it could obtain the sensing data from a neighbor. The CE-CE interface shows how a CE would provide data to and receive data from a data archive. The S-S link is used to transfer sensing data between sensors. Typically, this would occur when one sensor is aggregating a number of measurements into a single fused measurement.

The 1900.6 specification fully defines the logical interfaces, message structure, and data types for all sensor data that can be exchanged. Figure 7.13 [10] shows an example the interfaces used when exchanging data.

From Figure 7.13 [10], there are three primary interfaces defined by the 1900.6: the A-SAP, the M-SAP, and the C-SAP. In IEEE 1900 parlance, the **service access point** (SAP) is a logical interface to a service which is defined in one of the standards. In this case, the **communications SAP** (C-SAP) defines how a lower layer

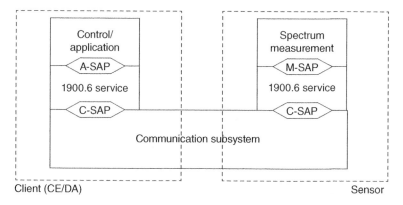

Figure 7.13 IEEE 1900.6 communication interfaces. *Source*: IEEE Std 1900.6-2011 [10].

communication system would need to format messages for a 1900.6 compliant node. Similarly, the **application SAP** (A-SAP) and **measurement SAP** (M-SAP) define how a component would talk to the 1900.6 interface on an application or sensor, respectively. Each of the messages for transferring data types and sending control commands to the applications and sensors is defined in the 1900.6 specification.

In addition to defining the interfaces and messages sent between the entities in the sensing system, the 1900.6 specification goes further and specifies what data can be sensed. Table 6.2 in Section 6.3 of the standard [10] provides a comprehensive listing of the data that can be exchanged between sensors and CEs in the network. The list contains a multiple information types in addition to the sensed spectrum data. Some examples include information about the sensor and its capabilities, policy requirements being enforced, accuracy estimates for measurements, network information, and many others.

The first version of 1900.6 was published in 2011 [10] and a corollary with minor changes was 2015 [19]. In 2014, an amendment, 1900.6a [20], was released which includes enhanced specifications for data archives.

7.1.8   1900.7 Working Group

The 1900.7 working group, "Radio Interface for White Space Dynamic Spectrum Access Radio Systems Supporting Fixed and Mobile Operation" presents both MAC and PHY layer specifications for radios operating within white space bands. Up until this point, the DySPAN standards committee had refrained from specifying specific transmitter and receiver protocol specifications, focusing instead on providing common lexicon surrounding spectrum sharing and methodologies for coexistence between different networks.

As *television white space* (TVWS) regulations were solidifying around the world, the DySPAN-SC and different standards were being developed, the SC determined that the current standards did not meet the needs of all potential TVWS band users, as they were primarily focused on supporting regional wireless internet providers and wide area "internet of things" networks [21]. The working group determined that the TVWS bands would provide good propagation characteristics and regulatory frameworks for some additional use cases including wireless backbones, ad hoc maritime networks, and home/office networking. The interfaces defined in 1900.7, therefore, are focused on supporting this wider range of use cases [22].

The specification is designed to be agnostic to the primary user of the channel, node mobility, and transmitter power. However, the most common implementation is expected to provide fixed wireless access to consumers in the TVWS allocations created by several countries during the broadcast television transition from analog to digital waveforms.

Figure 7.14 [11] shows the reference model for a 1900.7 device. The document specifies the function of the two shaded boxes (the MAC sublayer and PHY layer) as well as the interfaces between the various layer management entities.

The device's decisions about how and when to transmit are controlled by the three management entities and are specific to the device implementation. It seems likely that the management entities should comply with the related IEEE1900 standards.

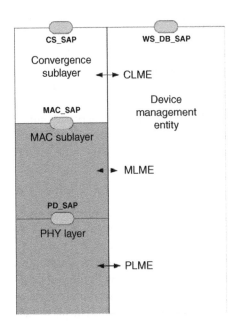

Figure 7.14 IEEE 1900.7 device reference architecture. *Source*: IEEE Std 1900.7-2015 [11].

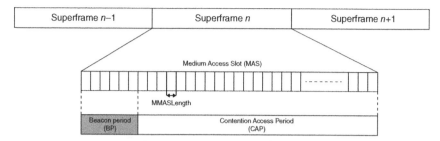

Figure 7.15 IEEE 1900.7 frame and superframe structure. *Source*: IEEE Std 1900.7-2015 [11].

For example, the *policy enforcer* from 1900.5 could likely be performing the role of all three-layer management entities if a policy-based reasoning system is built on top of the 1900.7 MAC/PHY layers.

The MAC sublayer provides two services: it interfaces with the **MAC Layer Management** entity (MLME) to determine the optimal transmission parameters and it formats MAC protocol data units from the convergence layer for the PHY layer. The media access schema is a slotted *carrier-sense multiple access with collision avoidance* (CSMA/CA) architecture with frame/superframe boundaries as shown in Figure 7.15. Each **superframe** begins with a configurable number of slots reserved for beacon frames. **Beacon frames** are sent by the master node in a network to provide information about the frame structure, network layout, and potential primary channel users.

1900.7 networks are able to form in an ad-hoc manner. When a device first scans the spectrum allocation looking for networks, it may find none. In this case, the node will begin transmitting superframes using the first beacon slot and wait for other nodes to join the network. If there are networks transmitting beacon frames, but not a network suitable to the device, it will begin transmitting beacons in the next available slot. If the device finds a beacon frame from a suitable network, it will attempt to join that network with an association request and, when completed, can send begin sending and receiving data.

The PHY layer provides two services as well: interfacing with the **PHY Layer Management Entity** (PLME) to determine transmission parameters and transmitting frames received from the MAC sublayer at the correct time.

Figure 7.16 [11] shows an example of the PHY layer symbol mapping which consists of a **filter bank multicarrier** (FMBC) symbol with a preamble containing a combination of null carriers (white), evenly spaced type I non-null carriers (Dark Gray), and sparsely distributed type II non-null carriers (black). Following the preamble, the data stream is distributed into the FMBC symbol index columns depending on the mode. The data stream consists of the payload from the MAC Layer, after a 16-bit cyclic redundancy check (CRC) has been appended, which gets scrambled,

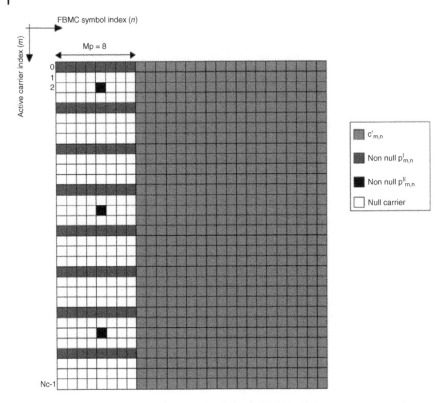

Figure 7.16 IEEE 1900.7 FMBC symbol mapping example. *Source*: IEEE Std 1900.7-2015 [11].

modulated, and coded with a punctured convolutional code. The PLME can vary the modulation type and coding rate based on input from the rest of the system.

The initial 1900.7 specification was finalized in 2015 [11]. However, at the time of writing, the authors were not able to find any commercially available instances of a 1900.7 radio on the market.

References

1 Harada, H., Alemseged, Y., Filin, S. et al. (2013). IEEE dynamic spectrum access networks standards committee. *IEEE Communications Magazine* **51** (3): 104–111.

2 Granelli, F., Pawelczak, P., Prasad, R.V. et al. (2010). Standardization and research in cognitive and dynamic spectrum access networks: IEEE SCC41 efforts and other activities. *IEEE Communications Magazine* **48** (1): 71–79.

3 Buljore, S., Harada, H., Filin, S. et al. (2009). Architecture and enablers for optimized radio resource usage in heterogeneous wireless access networks: The IEEE 1900.4 Working Group. *IEEE Communications Magazine* **47** (1): 122–129.

4 Sherman, M., Mody, A.N., Martinez, R. et al. (2008). IEEE standards supporting cognitive radio and networks, dynamic spectrum access, and coexistence. *IEEE Communications Magazine* **46** (7): 72–79.

5 Prasad, R.V., Pawelczak, P., Hoffmeyer, J.A., and Berger, H.S. (2008). Cognitive functionality in next generation wireless networks: standardization efforts. *IEEE Communications Magazine* **46** (4): 72–78.

6 IEEE Std 1900.1-2008. (2008). IEEE Standard Definitions and Concepts for Dynamic Spectrum Access: Terminology Relating to Emerging Wireless Networks, System Functionality, and Spectrum Management, IEEE.

7 I. 1. S. Committee. (1998). IEEE 1900.2, IEEE.

8 I. 1. S. Committee. (2011). IEEE 1900.4, IEEE.

9 I. 1. S. Committee. (2011). IEEE 1900.5, IEEE.

10 IEEE Std 1900.6-2011. (2011). IEEE Standard for Spectrum Sensing Interfaces and Data Structures for Dynamic Spectrum Access and Other Advanced Radio Communication Systems, IEEE.

11 IEEE Std 1900.7-2015. (2015). IEEE Standard for Radio Interface for White Space Dynamic Spectrum Access Radio Systems Supporting Fixed and Mobile Operation, IEEE.

12 IEEE Std 1900.1-2019 (2019). IEEE Standard for Definitions and Concepts for Dynamic Spectrum Access: Terminology Relating to Emerging Wireless Networks, System Functionality, and Spectrum Management. IEEE Std 1900.1-2019 (Revision of IEEE Std 1900.1-2008), pp. 1–78, doi: https://doi.org/10.1109/IEEESTD.2019.8694195.

13 Filin, S., Harada, H., Murakami, H. et al. (2009). IEEE 1900.4 generic procedures to realize dynamic spectrum access use cases. *IEEE EUROCON*, St. Petersburg.

14 IEEE. (2001). 1900 Standards Committee, IEEE 1900.4a, IEEE.

15 IEEE 1900 Standards Committee. (2013). IEEE 1900.4.1, IEEE.

16 IEEE 1900 Standards Committee. (2017). IEEE 1900.5.2, IEEE.

17 Noguet, D. (2009). Sensing techniques for Cognitive Radio – State of the art and trends, SCC41 – P1900.6.

18 Murroni, M., Prasad, R.V., Marques, P. et al. (2011). IEEE 1900.6: spectrum sensing interfaces and data structures for dynamic spectrum access and other advanced radio communication systems standard: technical aspects and future outlook. *IEEE Communications Magazine* **49** (12): 118–127.

19 IEEE Std 1900.6-2011/Cor 1-2015 (Corrigendum to IEEE Std 1900.6-2011). (2016). IEEE Standard for Spectrum Sensing Interfaces and Data Structures for Dynamic Spectrum Access and Other Advanced Radio Communication Systems – Corrigendum 1, IEEE.

20 I. 1. S. Committee. (2014). IEEE 1900.6a, IEEE.

21 Hoang, F.H. (2012). Introduction to 1900.7 network architecture, design and current status. *IEEE International Conference on Networks (ICON)*.

22 Filin, S., Noguet, D., Dore, J.-B. et al. (2018). IEEE 1900.7 standard for white space dynamic spectrum access radio systems. *IEEE Communications Magazine* **56** (1): 188–192.

8

Coexistence Standards in IEEE 802

The goal of this chapter is to provide an overview of various coexistence standards and identify important coexistence mechanisms within them. Many of the standards addressed have components outside the context of coexistence. Some standards have sections that have never been deployed. The coexistence mechanisms highlighted have been curated such that the mechanisms are in active use.

Earlier chapters have provided in-depth analysis of the coexistence mechanisms highlighted here. An extensive list of reference material is provided at the end of those chapters for the interested reader. This chapter will provide the reader with details on how those concepts are employed in existing wireless coexistence standards.

8.1 The Standards to Be Addressed in this Chapter

There are many wireless standards which seek to provide a means of wireless coexistence. One of the first questions that may come to mind is: Why so many?

The different standards for wireless coexistence address different sets of problems. Wireless coexistence standards for operation in *TV White Space* (TVWS) need to provide deference to the TV broadcasters. This demonstrates that wireless coexistence standards must contend with legacy spectrum regulation. Wireless coexistence in the Industrial, Scientific, and Medical (ISM) 2.4 GHz band needs to address interference. Thus, the different standards exist because the authors were focused on different subsets of the wireless coexistence problem space.

Industry groups that organize to develop standards for operations in different bands often coordinate with independent standards bodies, such as the IEEE. Such a partnership allows the standard to be maintained by a neutral party;

Wireless Coexistence: Standards, Challenges, and Intelligent Solutions, First Edition.
Daniel Chew, Andrew L. Adams, and Jason Uher.
© 2021 The Institute of Electrical and Electronics Engineers, Inc.
Published 2021 by John Wiley & Sons, Inc.

that neutrality provides assurance of fair play, and that encourages new developers to use the standard.

This book will focus on multiple wireless standards for coexistence. Those are:

- IEEE 802.22
- IEEE 802.19.1
- IEEE 802.15.2 and the ISM Band
- IEEE 802.11af

As discussed in Chapter 1, there are five categories of coexistence strategies. Those are **Separation, Mitigation, Monitoring, Sensing,** and **Collaborative.** A wireless coexistence standard may employ multiple strategies as will be seen.

8.2 Types and Spatial Scope of Wireless Networks

Considerations made in defining that standard are a direct product of the geographic range to be covered. This chapter will cover standards intended for small Personal Area Networks and very large Regional Area Networks.

A comparison of wireless networks based on *spatial scope* comes from reference [1] and is shown in Figure 8.1. In this version of network comparison by spatial scope, individual networks defined by the IEEE 802 family of standards are identified. Also identified are the frequency ranges in which those standards operate. The propagation characteristics of these frequency ranges will impose limits on the potential range of the various networks.

8.3 Stacks: The Structure of Wireless Protocol Standards

The seven-layer **Open Systems Interconnection (OSI) Reference Model** defined by the International Standardization Organization (ISO) [2] is the academic standard when analyzing the protocol stack of a given system. Interestingly it is not the most popular in practice. The **TCP/IP Reference Model**, defined by the Internet Engineering Task Force (IETF) RFC1122 [3], is the protocol stack which defines the Internet. These two stacks are compared side by side in Figure 8.2. The three upper layers of the OSI stack are condensed into one Application layer in the TCP/IP model. The two lower layers of the OSI model were condensed into the TCP/IP Link layer. The TCP/IP reference model is the simpler of the two and wound up being the one chosen to implement what we now know as

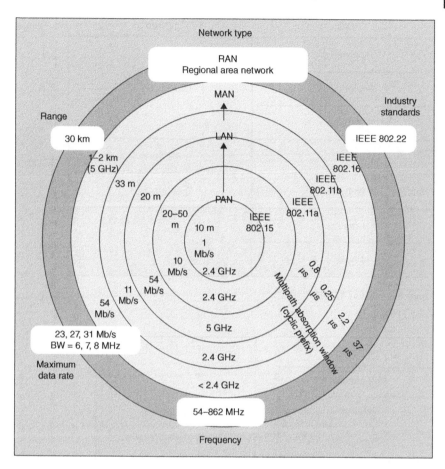

Figure 8.1 Network types and sizes. *Source:* Stevenson et al. [1]. (© [2020] IEEE).

the Internet [4, 5]. References [4, 5] provide a history of the OSI model and TCP/IP model and how TCP/IP won out.

In many of the standards discussed in this chapter there is a comparison of that standard's protocol stack to the OSI Reference Model. The IEEE 802 standards follow this OSI model. The Data Link layer of the OSI stack can be broken into two sublayers, the Media Access Control layer and the Logical Link layer. Most IEEE 802 standards only define the Physical and Media Access Control layers. IEEE 802.2 provides a common Logical Link layer. A relation between the IEEE 802 standards and the OSI Reference Model is shown in Figure 8.3.

The IEEE 802 standards often describe how they fit into the OSI model as an academic exercise. However, it does not get nodes connected to the internet.

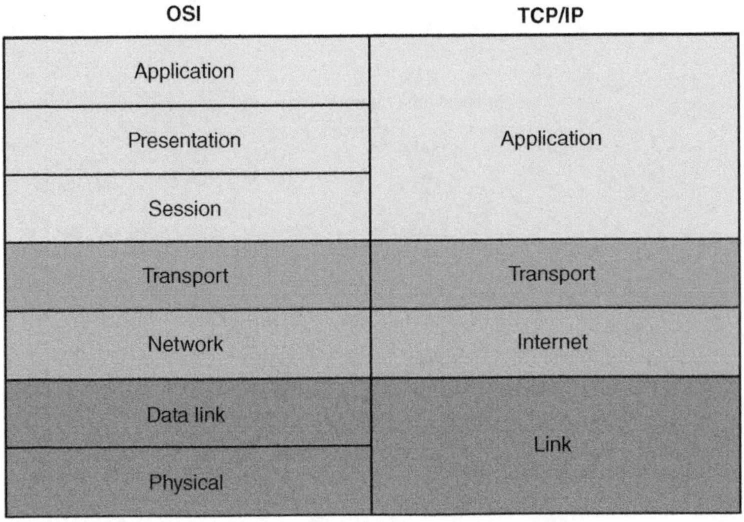

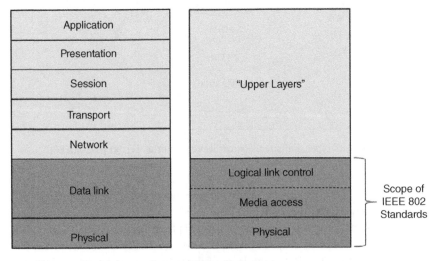

Figure 8.2 OSI vs TCP/IP.

Figure 8.3 IEEE 802 standards in the OSI model.

For that purpose, Figure 8.4 has been provided. Figure 8.4 shows how the IEEE 802 standards fit with the TCP/IP model. The two examples shown are Wi-Fi (802.11) and Ethernet (802.1). Those two IEEE 802 standards converge with IEEE 802.2 which provides a common Logical Link layer. After IEEE 802.2, IETF RFC1122

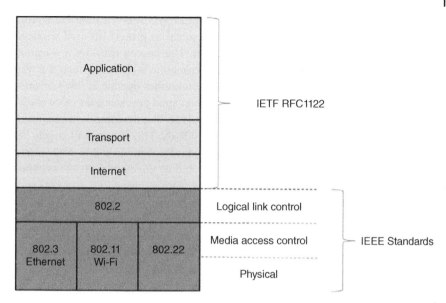

Figure 8.4 IEEE 802 Standards and TCP/IP.

defines the stack. Many other protocols defined in IEEE 802, such as IEEE 802.22 and IEEE 802.16, also fit into the paradigm illustrated in Figure 8.4.

In the following sections, various standards will be compared to the OSI model discussed above. Some standards provide data connectivity and will fit into the TCP/IP model, as shown in Figure 8.4.

8.4 IEEE 802.22

In 2002, the U.S. Federal Communications Commission (FCC) performed a study and found much of the allocated spectrum was being underutilized [6]. This underutilization opens up the possibility of *spectrum sharing*. In 2004, the FCC issued a proposed change to the rules of spectrum use in the bands otherwise allocated to television stations [7]. The IEEE 802.22 working group (IEEE 802.22 WG) was formed in direct response to this proposal for the use of *Television White Spaces* (TVWS) in the United States. The IEEE 802.22 WG began working on a standard to provide a **Wireless Regional Area Network** (WRAN) service using TVWS. This work began before the FCC issued rules for access to TVWS. The working group began to publish papers on the subject as early as 2006 [8] touting the coming IEEE 802.22 standards as the first wireless standard for cognitive radios. It was not until 2008 that the FCC published rules for accessing this White Space [9]. In 2009, the National Association of Broadcasters filed suit to prevent the rollout of TV white space access rules.

The first standard produced by the IEEE 802.22 WG was IEEE 802.22.1 [10], published in 2010, which provided a beacon signal to protect licensed wireless microphones operating in the television bands. The beacon provides a warning to WRAN systems that a licensed wireless microphone is operating near a given television frequency-channel. The wireless microphones operate at low transmit power and may be difficult to detect. The beacon signal provides a far more easily detected warning that an incumbent is present [11]. The next standard produced was IEEE 802.22-2011 [12] which defined the WRAN. This section will largely be dedicated to that WRAN standard. A third standard was published in 2012 which provides recommendations for the installation and deployment of 802.22 systems [13]. Two amendments to the WRAN standard were published in 2014 [14] and 2015 [15]. A revision to the 2011 standard was published in May of 2020, IEEE 802.22-2019 [16]. IEEE 802.22-2019 rolled up the enhancements in the prior amendments and expanded the scope of 802.22 from TV white space to opportunistic spectrum access in multiple bands. IEEE 802.22 is the standard for the WRAN. The *WhiteSpace Alliance* does interoperability testing.

IEEE 802.22 [16] is a WRAN standard developed to facilitate white space access in the television band. The intended use-case is providing data connectivity to rural areas. IEEE 802.22 seeks to achieve this through dynamic spectrum access in the white spaces in the television band. The dynamic spectrum access must provide protection to all *incumbents*, those being licensed users of the television band. These incumbents include the television signals themselves and wireless microphones licensed for operation in that band. Incumbent signals in the TV Band are the *Primary User*. The IEEE 802.22 WRAN is the *Secondary User*.

The primary purpose of IEEE 802.22 is to provide data connectivity in rural environments using unused spectrum found in *TV White Space*. IEEE 802.22 has been touted as one of the world's first cognitive radio standards [1]; however, the standard does not exclusively deal with this issue. A significant portion of the standard defines a data plane by which data connectivity can be delivered.

The IEEE 802.22 WRAN standard defines a network that is point-to-multipoint where a base station (BS) supports multiple subscriber nodes. The individual subscriber nodes are located with each individual customer and are called Customer Premises Equipment (CPE). The base station is in control of the network. The standard refers to the network formed by CPEs and a BS as a *WRAN cell*. The terms *WRAN* and *cell* are also used by themselves in the standard to refer to the network, and all three are used interchangeably. The base station doles out time-frequency resources to the CPEs, performs spectrum sensing, provides an interface to the Wide Area Network, and manages quality of service for the different CPEs.

The IEEE 802.22 WRAN employs spectrum sensing and a database of known incumbent emitters to determine which frequency channels are available for use. The operation of the WRAN is illustrated in Figure 8.5 which comes from reference [17]. The incumbent is the TV Transmitter and the Wireless Microphones

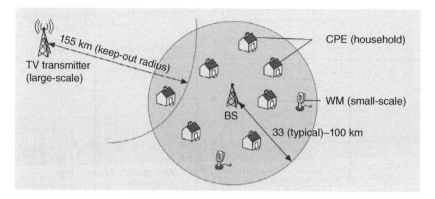

Figure 8.5 IEEE 802.22 WRAN operation. *Source*: Shin et al. [17].

(WM). An area around the TV transmitter is determined to be not valid for WRAN use and shown with the "keep-out" radius. Spectrum sensing alone was found to be insufficient in this application to protect the incumbent signals. The required geolocation database can inform the WRAN of the presence of the TV Station and the WRAN can then enforce the keep-out distance. The spectrum-sensing feature of the WRAN may detect the microphones in operation or a beacon signal can warn the WRAN of the existence of the microphones.

The WRAN standard defines a protocol stack (referred to as the *protocol reference model* in the standard) with three planes. Those three planes are:

- The **data plane** defines the physical (PHY) and medium access control (MAC) layers of the wireless communications link. This link provides the wireless connectivity between the BS and CPE. The data plane follows the standard protocol stack configuration in Figure 8.6. The MAC layer takes direction from the cognitive plane and control plane.
- The **control plane** contains the Management Information Base (MIB). An MIB is a common term for a networking management database. The devices in an 802.22 network are managed using the Simple Network Management Protocol (SNMP). The management plane also connects to an external database of known emitters.
- The **cognitive plane** provides *sensing information* to the other planes. The cognitive plane is divided into two layers. The lower layer handles geolocation and the spectrum sensing functions, providing that information to the upper layer. The upper layer contains the *Spectrum Manager* (SM) and *Spectrum Sensing Automation* (SSA), in the case of the base station (BS), or only the SSA in the case of customer premises equipment (CPE). The SSA is a simplified version of the SM. The SSA has limited roles and only operates with CPE that have

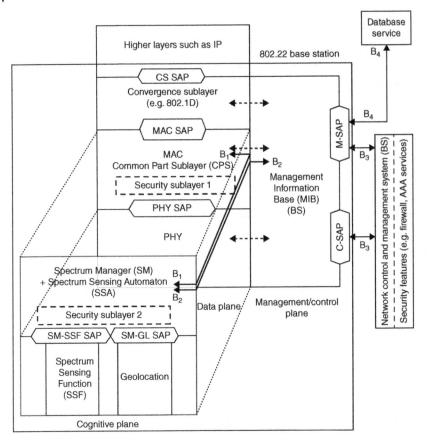

Figure 8.6 The IEEE 802.22 protocol reference model of the BS. *Source*: IEEE 802. 22-2019 [16].

not yet associated with the base station of a WRAN cell. Once the CPE has established association with the base station of the WRAN cell, the SAA ceases operations and the SM of the Base Station assumes all control of the cognitive plane for both BS and CPE. The upper layer interfaces with the other planes, taking in information to make decisions on spectrum availability, and providing information to the data plane.

The IEEE 802.22 provides two illustrations of the protocol stack, called the *protocol reference model* in the standard. One of the illustrations is for a base station and the other is for CPE. The protocol stack for the base station as illustrated in the standard is shown in Figure 8.6. It is a three-dimensional visualization, connecting the three planes. The interfaces between layers are defined by *Service Access Points*

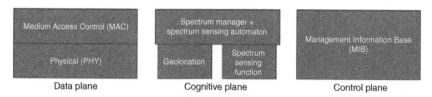

Figure 8.7 Simplified view of the IEEE 802.22 protocol stack.

(SAP). The cognitive plane is connected to the data plane at the MAC layer. Both layers of the data plane are connected to the control plane. The MAC layer is shown to have two sublayers, the *Convergence Sublayer* and the *Common Part Sublayer*.

Figure 8.7 shows a simplified, flattened, view of this protocol stack. The cognitive plane shows a layer akin to MAC which contains either the *Spectrum Manager* (SM) in the case of the BS or *Spectrum Sensing Automaton* (SSA) in the case of CPE. This SM-SAA layer then connects to two lower layers, those being the *Spectrum Sensing Function* (SSF) and *Geolocation*. The SM of the cognitive plane is connected to the control plane. The structure shown in Figure 8.7 applies to both the CPE and the BS; however, the SM only exists in the BS. The upper layer of the cognitive plane for the CPE consists of the SSA alone. Figure 8.7 does not show the interconnects. In order to simplify the stack, these interconnects will be shown in subsequent stack illustrations at the start of every section describing the operation of one of the three planes.

Each of these three planes will be discussed in the following sections, starting with the data plane. The sensing techniques used in the cognitive plane will also be discussed. The theory behind these sensing techniques will be discussed in detail in earlier chapters. For the purpose of describing the 802.22 standard, these sensing techniques will be summarized.

8.4.1 The Data Plane

Figure 8.8 shows the simplified stack from Figure 8.7 redrawn to place the data plane in the center. The MAC layer is shown to have two sublayers, the *Convergence Sublayer* and the *Common Part Sublayer*. The connections between the data plane and the other two planes are shown. The *Spectrum Manager* and *Spectrum Sensing Automaton* from the cognitive plane connect to the Common Part Sublayer of the MAC. The control plane connects to the MAC layer as a whole. The control plane also connects to the PHY layer.

IEEE 802.22 is an Orthogonal Frequency Division Multiple Access (OFDMA) scheme. Orthogonal Frequency Division Multiplexing (OFDM) and OFDMA are described in more detail in Chapters 3 and 4, respectively. As a brief overview,

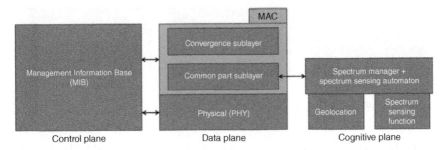

Figure 8.8 Simplified view of the IEEE 802.22 protocol stack, centered on the data plane.

OFDMA is a multiple access scheme in which all nodes in network use OFDM symbols. OFDM symbols are a "multicarrier" modulation scheme in which multiple orthogonally located subcarriers are linearly modulated.

IEEE 802.22 uses adaptive coding and modulation (ACM). This means that the underlying subcarrier modulation is dynamically selected by monitoring the quality of the wireless link. In addition to dynamically setting the modulation order, the forward error-correction coding is also determined dynamically.

Figure 8.9 from reference [12] illustrates the IEEE 802.22 adaptive modulation as a function of range. Wireless links of CPEs close to the base station suffer less propagation loss and the wireless channel can afford a higher modulation order, in this case 64-QAM. As the CPEs move farther away, the received signal strength is

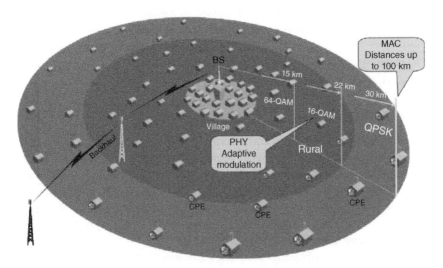

Figure 8.9 IEEE 802.22 adaptive modulation as a function of range. *Source*: IEEE 802. 22-2019 [16]

diminished. In order to maintain a wireless link, the modulation order must decrease to 16-QAM. Finally, at the furthest geographic reaches of the network, the modulation order must decrease to QPSK in order to maintain the wireless link.

8.4.2 The Control Plane

Figure 8.10 shows the simplified stack from Figure 8.7 redrawn to place the control plane in the center. The connections between the control plane and the other two planes are shown. The control plane connects to the MAC layer of the data plane as a whole. The control plane connects to the *Spectrum Manager* and *Spectrum Sensing Automaton* layer of the cognitive plane.

The control plane contains the Management Information Base (MIB). The concept of a MIB appears in numerous network standards. A MIB is hierarchical database used for managing networks. The MIB contains a record for all managed devices in the network. The Simple Network Management Protocol (SNMP) [18] is used to manage devices in the network. The SNMP standard uses the term "MIB" to describe the database of managed devices.

In addition to containing the MIB, the control plane interfaces with a database of known incumbent signals. This database of known incumbent signals is referred to in some literature as the *geolocation database*. The IEEE standard refers to the *database service* as part of its protocol stack, and also refers to an *incumbent database*. The term "incumbent database" is synonymous with "geolocation database" as both concepts are meant to describe a database with records of incumbent signals listed by geographic location. The term "database service" is meant to describe the feature as a whole. The IEEE 802.22 standard does not specify how the geolocation database feature is to be implemented. One possible choice is to use the Internet Engineering Task Force (IETF) Protocol to Access White Space (PAWS) Databases standard [19]. In the United States, the FCC approves geolocation database administrators. The databases must be publicly accessible. Further information on these databases is discussed in the commercialization section.

Figure 8.10 Simplified view of the IEEE 802.22 protocol stack, centered on the control plane.

8.4.3 The Cognitive Plane

Figure 8.11 shows the simplified stack from Figure 8.7 redrawn to show the connections between the cognitive plane and the other two planes. The *Spectrum Manager* (SM) and *Spectrum Sensing Automaton* (SSA) connect to the Common Part Sublayer of the MAC layer from the data plane. The control plane connects to the SM and SSA. *Geolocation* and the *Spectrum Sensing Function* are on the same level as the PHY in the data plane. The SM and SSA take the role of the MAC layer.

In addition to monitoring a wireless link and adapting to the modulation in response to channel impairments, the IEEE 802.22 standard employs spectrum sensing to access white space in the television bands. The cognitive plane is in charge of spectrum sensing and geolocation. An example of spectrum sensing in IEEE 802.22 is illustrated in Figure 8.12 from reference [16]. Figure 8.12 shows an arbitrary spectrum with frequency as the x-axis. Incumbent emitters use certain frequency-channels. Those frequency-channels must be avoided by the WRAN. Figure 8.12 shows multiple 802.22 WRAN system operating concurrently.

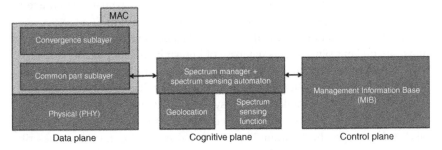

Figure 8.11 Simplified view of the IEEE 802.22 protocol stack, centered on the cognitive plane.

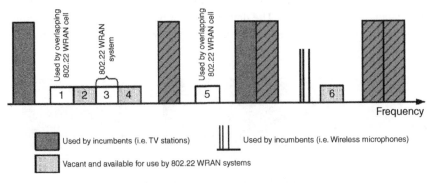

Figure 8.12 IEEE 802.22 spectrum sensing. *Source*: IEEE 802.22-2019 [16].

8.4.3.1 The Spectrum Manager

The **Spectrum Manager** (SM) is the central component of the spectrum sensing feature in the IEEE 802.22 standard. The SM is illustrated in Figure 8.13 from reference [12] and has seven subordinate components: *CPE Registration and Tracking, Policies, Self-Coexistence, Spectrum Sensing Automations, Incumbent Database Service, Geolocation*, and *Channel Set Management*.

8.4.3.2 The Spectrum Manager Components

The **CPE Registration and Tracking** component refers to the need for the SM to consider the capabilities of Customer Premise Equipment (CPE), availability of channels, requirements of local incumbents, and other restrictions before allowing CPE to associate with the WRAN cell.

The **Policies** component represents the responsibility of the SM to enforce certain rules within the cell. Those rules are referred to as *policies* in the standard and the standard provides a list. The list describes specific events and what actions to take if those events occur.

The **Self-Coexistence** component enables coordination with other neighboring IEEE 802.22 WRANs. The self-coexistence component of the spectrum manager provides a beacon to other 802.22 WRAN systems so as to increase the likelihood of two 802.22 WRANs systems detecting one another. Additionally, it allows two 802.22 WRAN systems to coexist in the same frequency-channel by implementing a time-division multiplexing access scheme for the singular frequency-channel. This prevents neighboring 802.22 WRANs from using the same frequency-channel at the same time.

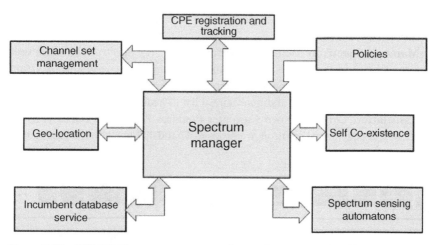

Figure 8.13 IEEE 802.22 spectrum manager. *Source*: IEEE 802.22-2019 [16].

The **Spectrum Sensing Automations** perform spectrum-sensing functions. The spectrum sensing function is performed with a look-through window. This look-through window is called "a quiet period" in the standard. The idea behind the quiet periods is that any attempt to sense the spectrum while the system is transmitting would result in a self-detect. The system cannot confirm that a given frequency-channel is unused while it is using that resource itself. There are two types of quiet periods defined in the standard resulting in what the standard calls the "two-stage sensing mechanism". This is illustrated in Figure 8.14 from reference [12]. The *intra-frame sensing* takes place once per frame. The duration of this sensing period is less than one frame. The *inter-frame sensing* exists if the system needs a sensing period longer than one frame size. The base station will determine if inter-frame sensing is needed.

The **Incumbent Database Service** component is the means by which the Spectrum Manager accesses the database of known incumbent emitters. If the local regulatory authority requires a database of incumbent signals, then the database will be operated under the rules of that regulatory authority. In the case that the regulatory authority does not require such a database, then the WRAN operator must provide a substitute.

The **Geolocation** component is the means by which the spectrum manager establishes the location of the base station. The spectrum manager also queries and maintains the locations of the Customer Premises Equipment.

The **Channel Set Management** component keeps track of the state of frequency channels. This involves *Channel Classification and Selection* as will be discussed in Section 8.4.3.3.

8.4.3.3 The Spectrum Manager Responsibilities

The *Spectrum Manager* (SM) components are used in the fulfillment of the spectrum manager duties.

Maintain Spectrum Availability Information: The Spectrum Manager uses information from the *Geo-location*, *Incumbent Database Service*, and *Spectrum Sensing Automations* components to determine the availability of frequency-channels. The SM then updates and maintains a list of available frequency-channels.

Association Control: When Customer Premises Equipment (CPE) joins an IEEE 802.22 cell, it must go through a process of handshaking with the base station

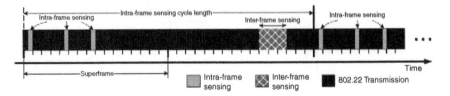

Figure 8.14 IEEE 802.22 two-stage mechanism. *Source*: IEEE 802.22-2019 [16].

(BS). The IEEE 802.22 standard refers to the entire process as *initialization and network association*. The SM is responsible for granting (or denying) association to requesting CPE. The SM must consider the location of the CPE, the capabilities of the CPE requesting association, and consider the impact of the CPE on incumbents.

Channel Classification and Selection: The SM categorizes all frequency-channels into the categories in Table 8.1. The SM assigns frequency channels in the WRAN cell. The SM determines on which frequency channel the WRAN cell will operate (the *operating channel*), designates *backup channels* in case the operating channel fails for any reason, and maintains a list of other frequency-channels with the status of that frequency channel.

Table 8.1 IEEE 802.22 channel categories.

Channel category	Description
Disallowed	**Disallowed Channels** are precluded from operation by the operator as opposed to being sensed as occupied or listed as protected in a database. The operator may configure the system to preclude channels, for example, in order to meet local regulations.
Operating	The **Operating Channel** is the frequency channel currently being used by the data plane in an IEEE 802.22 cell. This channel shall be sensed at least every 2 seconds. Specific signal types for sensing may be specified by the local regulations. Signal-specific sensing is discussed in *Sensing Techniques*.
Backup	A **Backup Channel** is a channel that the IEEE 802.22 cell can switch to immediately, if needed. There may be more than one backup channel.
Candidate	A **Candidate Channel** is a frequency channel that may become a Backup Channel. The Base station may request the CPEs to sense Candidate frequency-channels and report the results. Promoting a Candidate Channel to a Backup Channel depends upon this distributed sensing. The minimum requirement for a Candidate Channel to become a Backup Channel is that it be sensed as incumbent-free at least every 6 seconds for no less than 30 seconds.
Protected	A **Protected Channel** is one in which activity has been detected through the database service or sensing. A protected channel may become a candidate channel if the activity on that channel has ceased. A protected channel may become a backup channel if no activity has been detected on that channel. The minimum sensing period for this determination is 30 seconds using sensing intervals of 6 seconds.
Unclassified	An **Unclassified Channel** is a frequency channel that has not yet been sensed.

8.4.4 Distributed Sensing

Much like in the problems discussed in the CSMA section of Chapter 4, the Base Station cannot sense the environment at the location of the CPE. The CPE must sense the spectrum from its location and report to the Base Station the status of prospective channels. CPE associated with a Base Station of a WRAN cell will fall under the domain of the *Spectrum Manager* (SM) of that Base Station. That SM will coordinate all spectrum sensing and collect all results. Channel classifications will depend upon these results.

This is a case of **Distributed Sensing**. The duty of sensing the environment takes place at multiple disparate locations. For example, consider a system with a base station, a CPE, and a third-party emitter. This scenario is illustrated in Figure 8.15. The base station is far enough away from the third-party emitter that the base station cannot sense the presence of the third-party emitter. The CPE is much closer the third-party emitter. The CPE can sense the presence of the emitter and using the same frequency-channel will result in interference. The CPE will inform the Base Station of the presence of the third-party emitter. The Base Station will rely on both sensing from the CPE and itself in order to make spectrum allocation decisions.

8.4.5 Sensing Techniques

The IEEE 802.22 standard identifies several sensing techniques that may be used as implementations of the spectrum sensing function. The standard does not limit an 802.22 to these techniques. The standard supplies information on these techniques to aid in implementing the spectrum sensing function.

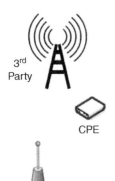

3rd Party

CPE

Base station

Figure 8.15 IEEE 802.22 distributed sensing scenario.

The spectrum sensing techniques discussed in the 802.22 standard are described as either *Blind* or *Signal Specific*, and as either *Coarse* or *Fine*. Both of those sets of qualifiers will define the spectrum sensing techniques. The qualifiers *Blind* and *Signal Specific* are mutually exclusive. The qualifiers *Coarse* and *Fine* are mutually exclusive.

A **Blind** sensing technique does not rely on any knowledge of the signal to be sensed. The terms "Blind" and "Non-Data Aided" are often used synonymously in other literature.

A **Signal-Specific** sensing technique does rely on knowledge of the signal to be sensed.

A **Coarse** sensing technique is intended to be run for a short period of time. The technique is not expected to reliably detect a signal at the required power levels but can detect high power signals in a short period of time.

A **Fine** sensing technique is only called such in relation to specific signals which can be guaranteed to be detected by the *Fine* method at the required power level. A *Fine* technique is intended to be run to make certain specific signals are detected. A *Fine* technique may be highly specialized to a specific signal. A *fine* technique is presumably more complex than a *Coarse* technique and may require samples over a longer period of time.

There are three Blind sensing techniques discussed in the 802.22 standard:

1) **Energy Detection**:
   Sums the energy of the received signal over some period and compares that energy to a threshold. The Energy detector is considered a *Coarse* technique for all signals.

2) **Eigenvalue Sensing**:
   Calculates estimates of the autocorrelation matrix of the received signal, finds the eigenvalues of that matrix, and then extracts test statistics from the eigenvalues. Eigenvalue Sensing is considered a *Fine* sensing technique for wireless microphones.

3) **Multi-Resolution Sensing**:
   Provides a coarse sensing. This technique uses a wavelet filter with a tunable center frequency and bandwidth. The filter is run over a range of center frequencies, and the power at each frequency is measured. The *whiteness* (flatness) of the spectrum is then tested. The IEEE 802.22 standard describes an analog implementation of this capability and indicates that this technique is intended for detecting Digital Television (DTV), making it a *Fine* sensing technique for DTV.

An 802.22 system may be required to detect specific signals. All signals that could possibly be required are listed in the IEEE 802.22 standard. If it is the case that the detection of certain signals is required, that list of required signals will be part of the configuration of the *Spectrum Manager*. Minimum sensitivity is also specified for any required signal detection.

The *Signal-Specific* sensing techniques are intended to be used when the 802.22 system is required to detect signals from a specified list. There are 11 *Signal-Specific* sensing techniques discussed in the 802.22 standard:

1) **ATSC Signature Sequence Correlation**
2) **ATSC FFT-Based Pilot Sensing**
3) **ATSC Pilot-Sensing Technique**
4) **ATSC PLL-Based Pilot Sensing**
5) **Wireless Microphone Covariance Sensing**

6) **ATSC Pilot Covariance Sensing**
7) **Spectral Correlation Sensing:**
   Note: this technique is intended to run signal-specific correlators in the frequency domain. This technique is therefore signal-specific, as it by definition requires a type of signal to be specified in order to run the correlation.
8) **ATSC Cyclostationary Sensing**
9) **Time-Domain Correlation Sensing for Analog Television**
10) **Improved Energy Detection for Analog Television**
11) **Time-Domain Correlation Sensing for DVB-T**

These signal-specific techniques exploit specific characteristics of these known signals to improve the probability of detection as a function of the *Signal-to-Noise Ratio* (SNR). The *Signal-Specific* techniques will be able to achieve higher probability of detection with lower probability of false alarms for a given SNR for the signal type against which the technique was designed. The signal-specific techniques may not work well at all against any other signal type.

8.5 IEEE 802.11

IEEE 802.11 is the wireless standard from which the popular "Wi-Fi" WLAN is defined. Wi-Fi has become a ubiquitous means of wireless connectivity. It is so ubiquitous as to be an amenity expected to be provided by businesses to their patrons. "Wi-Fi" chipsets come standard in an increasing number of consumer electronics [20]. "Wi-Fi" stands for "wireless fidelity" and it is not a standard but rather a trademark of the Wi-Fi Alliance. The Wi-Fi Alliance is a nonprofit trade organization responsible for managing device interoperability through certification [20, 21]. It is the Wi-Fi Alliance which manages the Wi-Fi technology branding. Use of the label "Wi-Fi" or "Wi-Fi CERTIFIED" means that a product meets the interoperability requirements established by the organization. The wireless protocol used for the "Wi-Fi" Wireless Local Area Network (WLAN) is defined by the IEEE 802.11 standard [22]. The Wi-Fi Alliance only certifies a subset of the waveforms contained in the standard. That is to say that although the IEEE 802.11 standard defines several waveforms, the link used as "Wi-Fi" only employs a subset of those waveforms. This book will treat the two terms, IEEE 802.11 and "Wi-Fi," as synonymous for the convenience of the reader.

The Wi-Fi specifications represent decades of standardization effort and are spread across thousands of pages. It is not the intention of this section to repeat that information, but rather to focus on issues which have practical importance to wireless coexistence. The Wi-Fi specification involves interference mitigation through the use of spread spectrum techniques, carrier sensing, and accessing *TV White Space*.

8.5.1 A Brief History of the IEEE 802.11 Standards

The IEEE 802.11 standards were first published in 1997 [23] as a wireless replacement for physical network cables in home and business computer network environments [21]. Success in the market and a drive to meet ever-growing demand has led to extensive research and development. This has resulted in numerous task groups being formed to add some functionality or capability to Wi-Fi. The task groups focus on adding an amendment to the standard. Lower case letters are selected in alphabetical order to identify that amendment. The Task groups are identified as "TG" and the lower-case letters represent the target amendment. This process yields terms such as 802.11a, 802.11b, 802.11g, and 802.11n. The letters l, o, q, and x are not used in order to prevent confusion with other 802 standards. Postfixes with two letters are now used, such as 802.11af, because the number of amendments has exhausted single letters. Many of the amendments introduce some new feature, but amendments can also be concerned with topics such as government regulations and coexistence recommendations.

The standards itself is updated every few years. For example, IEEE 802.11-2016 was published in 2016, and IEEE 802.11-2020 was published in 2021. The updates to the standard may roll-up some number of amendments into a single document. Incorporated amendments are retired; however, the amendment nomenclature is still used to refer to the capabilities. For example, the designation "IEEE 802.11b" is still used to refer to the enhancements to the DSSS physical layer introduced by that amendment, even though the amendment was formally incorporated into the IEEE 802.11-2007 specification.

8.5.2 The Evolution of Wi-Fi

The evolution of Wi-Fi is illustrated in Figure 8.16 which comes from reference [24]. At the base of Figure 8.16 along the horizontal axis is a timeline beginning in 1997 and continuing to 2021. The vertical axis illustrates the target operational bands, sub 1-GHz, 2.4 and 5 GHz ISM, and 60 GHz. The evolution of Wi-Fi begins with operations in the 2.4 GHz band with the standard "802.11". The standard offered two modes of *Spread Spectrum* for interference mitigation, *Frequency Hopping* (FHSS) and *Direct-Sequence* (DSSS), with a data rate of 2 Mbps. Only DSSS saw widespread commercial deployment.

The IEEE 802.11a amendment provided data rates up to 54 Mbps through the use of *Orthogonal Frequency Division Multiplexing* (OFDM) in the 5 GHz unlicensed band. The 802.11b amendment was developed in parallel to 802.11a.

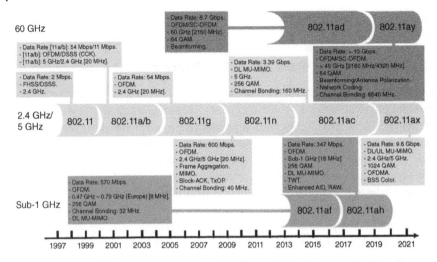

Figure 8.16 Evolution of the 802.11 Standard. *Source*: Selinis et al. [24], 1–1. open access article licensed under CC BY 4.0.

The DSSS physical layer was the basis for the enhancements introduced in the IEEE 802.11b amendment, which operated in the 2.4 GHz ISM band and increased the maximum data rate to 11 Mbps. The amendment 802.11g introduced OFDM to the 2.4 GHz ISM band at 54 Mbps. This OFDM capability was improved with 802.11n to offer up to 600 Mbps.

A few of the amendments are listed in Table 8.2 in a tabular format to help the reader with Figure 8.16. The amendments covered are 802.11a, 802.11b, 802.11g, 802.11n, 802.11ac, and 802.11af. Each of these amendments has been incorporated into the latest release of the standard, IEEE 802.11-2020.

8.5.3 Wi-Fi Channelization in the 2.4 GHz Band

Wi-Fi channels in the 2.4 GHz band have a bandwidth of 22 MHz. The channels are numbered starting with channel 1 at 2412 MHz and continuing to channel 13 at 2472 MHz. Depending on the regulatory rules of the country in which Wi-Fi is operating, there may be an additional channel 14 centered at 2484 MHz. This represents 5 MHz incremental steps between channels. Because the channels are 22 MHz wide and successive channels are only 5 MHz apart, the 2.4 GHz Wi-Fi channels overlap, as shown in Figure 8.17 [25]. Figure 8.18 [25] shows nonoverlapping channels. Co-located equipment can be set to such sets of channels, as shown in Figure 8.18, in order to avoid interference. One such set is channels 1, 6 and 11.

Table 8.2 Selected Wi-Fi Amendments.

Amendment	Status	Capability
a	Incorporated in IEEE 802.11-2007	Higher Speed PHY Extension in the 5 GHz Band
b	Incorporated in IEEE 802.11-2007	Higher Speed PHY Extension in the 2.4 GHz Band
g	Incorporated in IEEE 802.11-2007	OFDM in the 2.4 GHz Band
n	Incorporated in IEEE 802.11-2012	High Throughput (HT) OFDM PHY
ac	Incorporated in IEEE 802.11-2016	Very High Throughput (VHT) OFDM PHY
af	Incorporated in IEEE 802.11-2016	TV White Space Operations

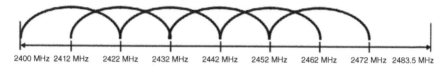

2400 MHz 2412 MHz   2422 MHz   2432 MHz   2442 MHz   2452 MHz   2462 MHz   2472 MHz 2483.5 MHz

Figure 8.17 Overlapping Wi-Fi channels. *Source*: IEEE 802.11b-199 [25].

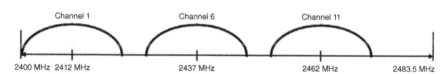

Channel 1   Channel 6   Channel 11

2400 MHz 2412 MHz   2437 MHz   2462 MHz   2483.5 MHz

Figure 8.18 Nonoverlapping Wi-Fi channels. *Source*: IEEE 802.11b-199 [25].

8.5.4 Carrier Sensing

Wi-Fi uses a *Carrier-Sense Multiple Access* (CSMA) scheme to enable multiple users on a given network. In this CSMA scheme, all stations in a given Wi-Fi network transmit and receive on the same frequency channel. Due to the use of only one frequency channel, CSMA depends on the ability of the Wi-Fi station to *sense* activity on that channel before transmitting. This is a concept called *Listen Before Talk*. The process of listening to the frequency channel before using it involves energy detection and detection of Wi-Fi signals based on known signal characteristics. CSMA was discussed in detail in Chapter 4.

8.5.5 Wi-Fi as TV White Space Access

IEEE 802.11af standard is an amendment to Wi-Fi that allows for dynamic spectrum access in *Television White Space* (TVWS). IEEE 802.11af makes use of geolocation databases for White Space access as do other TVWS standards. IEEE 802.11af is sometimes referred to as "White-Fi". The interest in using the TV Broadcast band stems from the superior propagation characteristics. Figure 8.19 from reference [26] (which gathered the data from NIST) shows the attenuation imparted on signals by common construction materials. The columns in Figure 8.19 categorize these attenuation values by operational frequency. As can be seen, the higher frequencies result in much higher attenuation in common materials.

802.11af amendment makes use of **GDD dependent stations** (dependent STAs) and **GDD enabling stations** (enabling STAs). GDD stands for *Geolocation Database Dependent*. The *enabling STA* serves like the traditional *Access Point*. The enablement procedure is shown in Figure 8.20 [27]. It is through this process that *dependent STAs* are enabled to use available TV channels. The *enabling STA* queries the **Geolocation Database** (GDB), which is labeled *TVWS database* in the figure. Upon getting results from the geolocation database, the *enabling STA* transmits a beacon. The *dependent STA* requests permission to transmit on a given open channel (White Space) from the *enabling STA*. Upon gaining permission, the *dependent STA* begins transmitting.

Materials	0.57 GHz (dB)	1 GHz (dB)	2 GHz (dB)	5.7 GHz (dB)	0.57 to 5.7 GHz (ΔdB)
Brick 89 mm	−1.5	−3.5	−5.4	−15	13.5
Brick 267 mm	−4.8	−7	−10.5	−38	33.2
Composite Brick 90 mm/ Concrete Wall 102 mm	−12	−14	−18	−42	30
Composite Brick 90 mm/ Concrete Wall 203 mm	−21.5	−25	−33	−71.5	50
Masonry 203 mm	−9.5	−11.5	−11	−12.75	3.25
Masonry 610 mm	−26.5	−27.5	−30	−46.5	20
Glass 6 mm	−0.4	−0.8	−1.4	−1.1	0.7
Glass 19 mm	−2.5	−3.1	−3.9	−0.4	−2.1
Plywood (dry) 6 mm	−0.15	−0.49	−0.9	−0.1	−0.05
Plywood (dry) 32 mm	−0.85	−1.4	−2	−0.9	0.05
Reinforced concrete 203 mm/ 1% steel	−23.5	−27.5	−31	−56.5	33
Reinforced concrete 203 mm/ 2% steel	−27.5	−30	−36.5	−60	32.5

Figure 8.19 Attenuation due to materials. *Source*: Flores et al. [26]. (© [2020] IEEE).

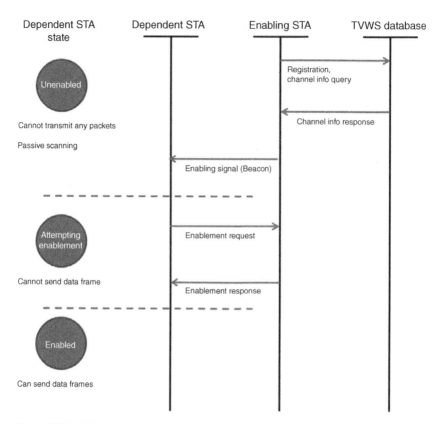

Figure 8.20 DSE processing procedure. *Source*: Kang et al. [27]. (© [2020] IEEE).

The 802.11af network is illustrated in Figure 8.21 [26]. The air interface between the two types of STA is defined by the 802.11af standard. The interface to the GDB is outside the 802.11af standard.

The channel bandwidths for 802.11af are defined by the "basic channel unit" (BCU) which may be 6, 7, or 8 MHz depending on the regulations of the region in which the system is operating. The BCUs may be combined together to form larger bandwidths. This concept is illustrated in Figure 8.22 from reference [26]. The physical layer of 802.11af is referred to as "TV High Throughput" (TVHT). TVHT is based on the VHT OFDM of 802.11ac.

8.5.6 Comparison of 802.11af and 802.22

There are many similarities between the wireless standards IEEE 802.22 and IEEE 802.11af as they are both designed to provide dynamic spectrum access in *TV White Space*. Both standards use a geolocation database. Both standards use OFDM symbols.

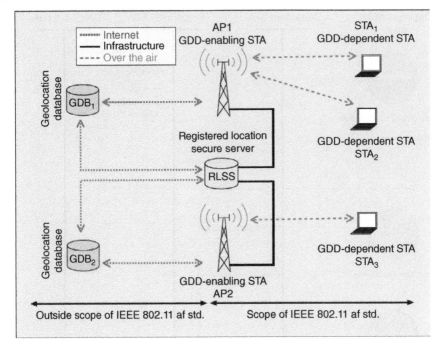

Figure 8.21 802.11af network. *Source*: Flores et al. [26]. (© [2020] IEEE).

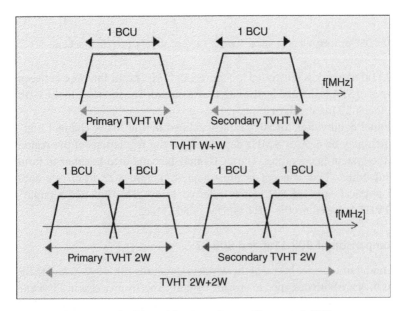

Figure 8.22 802.11af BCU combinations. *Source*: Flores et al. [26].

Key to the differences between the two standards is the respective intended use cases. IEEE 802.11af is intended to provide a WLAN whereas IEEE 802.22 is intended to provide a WRAN. As a WRAN, IEEE 802.22 is designed to cover longer distances than is IEEE 802.11af. Equipment costs for IEEE 802.22 are higher than that of IEEE 802.11af.

8.6 TVWS Geolocation Databases in the United States

Numerous companies applied to become TV white space database administrators when the FCC first made the opportunity available. Among the initial applicants were large corporations such as Goggle and Microsoft. Once the applicants were approved, they were required to be TVWS geolocation database administrators for a five-year term. Ten years after the FCC began taking applications, few of the original companies remain as TVWS geolocation database administrators. Many of the companies did their five-year term and left. This decreased interest indicates that operation in the *TV White Space* was not as lucrative as hoped.

8.7 IEEE 802.19.1

The wireless standards IEEE 802.22 and IEEE 802.11af are both designed to make use of *TV White Space*. The coexistence mechanisms in both standards are focused on homogeneous coexistence, that being coexistence between devices running the same wireless protocol. In the presence of a heterogeneous network, one in which devices of different wireless standards exist, the coexistence mechanisms defined by IEEE 802.22 and by IEEE 802.11af are not as effective.

IEEE 802.19 is the Wireless Coexistence Working Group (IEEE 802.19 WG) and it develops standards for wireless coexistence between different types of White Space radios. The first standard produced by the IEEE 802.19 WG was 802.19.1 in 2014 [28]. The original publication targeted operations in *Television White Spaces* (TVWS). The standard was revised in 2018 and broadened the scope of operations [29]. The IEEE 802.19.1 standard specified methods for wireless coexistence among dissimilar networks of unlicensed devices. This is distinct from IEEE 802.22. The IEEE 802.22 standard only specifies coexistence between incumbents (which are avoided) and other IEEE 802.22 nodes. For example, IEEE 802.22 provides a "self-coexistence mechanism". This does not extend to other secondary-user, nonincumbent, networks. Spectrum sensing may be able to help two secondary-user networks coexist. However, there is no direct coordination.

The IEEE 802.19.1 standard specifies a means of coexistence for nodes which do not otherwise share any standardization. The IEEE 802.19.1 standard assumes the unlicensed devices are capable of geolocation. The standard is not specific to any one band and may be applied to television white spaces, the 5 GHz license-exempt bands (5 GHZ ISM), and the general authorized access in the 3.5 GHz bands (CBRS).

The primary purpose of the IEEE 802.19.1 standard is coexistence. IEEE 802.19.1 does not deliver data connectivity to users. The standard instead de-conflicts access to white space in the spectrum. The IEEE 802.19.1 standard does not define a data plane or follow a traditional protocol stack. It does not interface with the Logical Link layer defined in IEEE 802.2. The IEEE 802.19.1 standard can be seen as providing an overlay onto the Medium Access Control layer of a registered White Space radio. This is because IEEE 802.19.1 influences the medium access decisions of that white space radio.

The architecture of the IEEE 802.19.1 standard is illustrated in Figure 8.23 from reference [30]. Figure 8.23 shows four types of nodes. The **WS Radio** node is a White Space radio such as an IEEE 802.22 WRAN. The WS Radio node connects to the **Coexistence Enabler** (CE) node. The CE node is a proxy between the WS Radio and the IEEE 802.19.1 coexistence system. The CE node connects to a **Coexistence Manager** (CM) node. The CM nodes connect to one **Coexistence Discovery and Information Server** (CDIS) Node.

The interface between a CE and a WS Radio node is not defined by the IEEE 802.19.1 standard. The CE node is intended as an application specific proxy and will be defined by the manufacturer of the WS Radio node.

White space radios register with CMs. Those radios provide the CMs with spectrum sensing information. The CM interfaces with the CDIS and a White Space database. The CMs send reconfiguration requests to registered white space radios based upon the information it has with regards to neighboring systems and available white space.

In addition to defining the system architecture, the IEEE 802.19.1 standard defines several algorithms for coexistence discovery and coexistence decisions. The coexistence discovery algorithms attempt to determine the impact of different

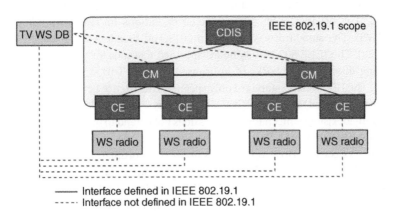

——— Interface defined in IEEE 802.19.1
----- Interface not defined in IEEE 802.19.1

Figure 8.23   IEEE 802.19.1 architecture.

White Space radios using conflicting time-frequency resources. If two White Space radios use the same frequency channel, or overlapping frequency channels, at the same time then there is risk of significant interference. The subsequent coexistence decision algorithms are then used to attempt to de-conflict the spectrum usage of the various White Space radios.

The interface to the White Space database is not defined by the IEEE 802.19.1 standard. An example implementation is the IETF PAWS protocol [19].

8.8   IEEE 802.15.2

IEEE 802.15.2 [31] is a set of recommendations, not a standard per se, and has been withdrawn. However, IEEE 802.15.2 is important to the history of the evolution of coexistence in the 2.4 GHz ISM band and is therefore worthy of some discussion.

Wi-Fi does not exist alone in the ISM band. Bluetooth is another popular waveform that operates in the 2.4 GHz ISM band. The evolution of Bluetooth was parallel to that of Wi-Fi. While Wi-Fi focused on WLAN connectivity, Bluetooth focused on WPANs.

Bluetooth was once based on an IEEE standard. In 1999, the Bluetooth SIG issued the specification for Bluetooth v1.0. The goal was to provide wireless connectivity for ancillary devices to cellphones. In 2002, the IEEE 802.15 committee approved the Wireless Personal Area Network (WPAN) standard 802.15.1 based on Bluetooth v1.0 [31]. The standard became IEEE 802.15.1-2002 [31].

As early as the 2000s, both Bluetooth and Wi-Fi reached a market large enough such that interference between the two became an active research topic. Bluetooth is a frequency-hopping protocol requiring a range of channels to hop across. That range spans 79 defined channels across the ISM Band. Therefore, Bluetooth will interfere with and suffer interference from Wi-Fi. References [33–35] are some examples of that research. Reference [33] provides analytical results of Wi-Fi (802.11b) packet error based on probability of packet collision. Reference [34] provides analytical results of Bluetooth packet error based on probability of packet collision. Reference [35] derives a closed-form solution for the probability of collision between the Bluetooth and Wi-Fi (802.11b). The interference between Bluetooth and Wi-Fi was of sufficient concern to warrant the creation of a task group to tackle it. The result of that task group was IEEE 802.15.2-2003 [31].

IEEE 802.15.2-2003 introduced ISM band coexistence mechanisms and recommendations. IEEE 802.15.2-2003 divided these recommendations into collaborative and noncollaborative mechanisms. Collaborative mechanisms required Bluetooth and Wi-Fi to communicate and coordinate. The noncollaborative methods could be implemented without any such coordination. Among the noncollaborative mechanisms recommended was *Adaptive Frequency-Hopping* (AFH). Frequency hopping and adaptive frequency hopping were discussed in detail in Chapter 4.

Bluetooth devices network together in a star topology. Such a Bluetooth network is called a *piconet*. The devices in a given *piconet* share a frequency-hopping pattern. The hopping pattern is a pseudo-random sequence of the 79 possible channels in the 2.4 GHz ISM band. AFH is used to exclude potential hop channels from the sequence of the entire *piconet* if that channel has exhibited poor throughput, presumably due to interference from a static (nonhopping) emitter.

AFH mitigated interference between Bluetooth and Wi-Fi, and thus provided for improved co-existence. The standard upon which Bluetooth was based at the time, IEEE 802.15.1-2002, was superseded by IEEE 802.15.1-2005. IEEE 802.15.1-2005 formally adopted improvements such as AFH [36].

After the 2005 revision of IEEE 802.15.1, IEEE no longer maintained the Bluetooth standard. The Bluetooth SIG took on the task of maintaining and evolving the standard. Those standards are called the Bluetooth Core Specifications, and version 5.1 was released in January 2019. Bluetooth continues to evolve and carries with it the AFH technique first put forward by IEEE 802.15.2.

IEEE 802.15.2 recommended other coexistence mechanisms in addition to adaptive frequency hopping; however, those mechanisms have not enjoyed adoption by one of the operating wireless standards.

References

1 Stevenson, C.R., Chouinard, G., Lei, Z. et al. (2009). IEEE 802.22: the first cognitive radio wireless regional area network standard. *IEEE Communications Magazine* **47** (1): 130–138.

2 Zimmerman, H. (1980). OSI reference model – the ISO model of architecture for open systems interconnection. *IEEE Transactions on Communications* **28** (4): 425–432.

3 Braden, R. (1989). RFC1122: Requirements for Internet hosts – communication layers, The Internet Engineering Task Force.

4 Russell, A.L. (2013). The internet that wasn't. *IEEE Spectrum* **50** (8): 39–43.

5 Meyer, D. and Zobrist, G. (1990). TCP/IP versus OSI. *IEEE Potentials* **9** (1): 16–19.

6 Federal Communication Commission. (2002). Spectrum Policy Task Force, Report ET Docket 02-135.

7 Federal Communication Commission. (2004). ET Docket 04-186, *Notice of Proposed Rulemaking*.

8 Cordeiro, C., Challapali, K., and Birru, D. (2006). IEEE 802.22: an introduction to the first wireless standard based on cognitive radios. *IEEE Journal of Communications* **1** (1): 38–47.

9 Federal Communication Commission. (2008). Second report and order and memorandum opinion and order in the matter of unlicensed operation in the television broadcast bands, *ET Docket No. 08-260*.

10 *IEEE Standard for Information Technology–Telecommunications and information exchange between systems–Local and metropolitan area networks–Specific requirements Part 22.1: Standard to Enhance Harmful Interference Protection for Low-Power Licensed Devices Operating in TV Broadcast Bands,* in IEEE Std 802. 22.1-2010, vol., no., pp.1–145, 1 Nov. 2010, doi: 10.1109/IEEESTD.2010.5623446.

11 G. J. Buchwald, S. L. Kuffner, L. M. Ecklund, M. Brown and E. H. Callaway Jr., *The Design and Operation of the IEEE 802.22.1 Disabling Beacon for the Protection of TV White space Incumbents,* 2008 3rd IEEE Symposium on New Frontiers in Dynamic Spectrum Access Networks, 2008, pp. 1–6, doi: 10.1109/DYSPAN.2008.55.

12 *IEEE Standard for Information technology– Local and metropolitan area networks– Specific requirements– Part 22: Cognitive Wireless RAN Medium Access Control (MAC) and Physical Layer (PHY) specifications: Policies and procedures for operation in the TV Bands,* in IEEE Std 802.22-2011, vol., no., pp.1–680, 1 July 2011, doi: 10.1109/IEEESTD.2011.5951707.

13 *IEEE Recommended Practice for Information Technology - Telecommunications and information exchange between systems Wireless Regional Area Networks (WRAN) - Specific requirements - Part 22.2: Installation and Deployment of IEEE 802.22 Systems,* in IEEE Std 802.22.2-2012, vol., no., pp.1–44, 28 Sept. 2012, doi: 10.1109/ IEEESTD.2012.6317267.

14 *IEEE Standard for Information Technology–Telecommunications and information exchange between systems Wireless Regional Area Networks (WRAN)–Specific requirements - Part 22: Cognitive Wireless RAN Medium Access Control (MAC) and Physical Layer (PHY) Specifications: Policies and Procedures for Operation in the TV Bands Amendment 1: Management and Control Plane Interfaces and Procedures and Enhancement to the Management Information Base (MIB),* in IEEE Std 802.22a-2014 (Amendment to IEEE Std 802.22-2011) , vol., no., pp.1–519, 30 May 2014, doi: 10.1109/IEEESTD.2014.6823051.

15 *IEEE Standard for Information Technology–Telecommunications and information exchange between systems - Wireless Regional Area Networks (WRAN)–Specific requirements - Part 22: Cognitive Wireless RAN Medium Access Control (MAC) and Physical Layer (PHY) Specifications:Policies and Procedures for Operation in the TV Bands - Amendment 2: Enhancement for Broadband Services and Monitoring Applications,* in IEEE Std 802.22b-2015 (Amendment to IEEE Std 802.22-2011 as amended by IEEE Std 802.22a-2014), vol., no., pp.1–299, 30 Oct. 2015, doi: 10.1109/ IEEESTD.2015.7336461.

16 *IEEE Standard - Information Technology-Telecommunications and information exchange between systems-Wireless Regional Area Networks-Specific requirements-Part 22: Cognitive Wireless RAN MAC and PHY specifications: Policies and Procedures for Operation in the Bands that Allow Spectrum Sharing where the Communications Devices May Opportunistically Operate in the Spectrum of Primary Service,* in IEEE Std 802.22-2019 (Revision of IEEE Std 802.22-2011), vol., no., pp.1–1465, 5 May 2020, doi: 10.1109/IEEESTD.2020.9086951.

17 Shin, K.G., Him, H., Min, A.W., and Kumar, A. (2010). Cognitive radios for dynamic spectrum access: from concept to reality. *IEEE Wireless Communications* **17** (6): 64–74.

18 Case, J., Fedor, M., Schoffstall, M. et al. (1990). RFC 1157: A Simple Network Management Protocol (SNMP), The Internet Engineering Task Force.

19 Chen, V., Zhu, D.S.L., and Malyar, J. (2015). RFC 7545: Protocol to Access White Space (PAWS) Databases, The Internet Engineering Task Force.

20 Hiertz, G.R., Denteneer, D., Stibor, L. et al. (2010). The IEEE 802.11 universe. *IEEE Communications Magazine* **48** (1): 62–70.

21 Ferro, E. and Potorti, F. (2005). Bluetooth and Wi-Fi wireless protocols: a survey and a comparison. *IEEE Wireless Communications* **12** (1): 12–26.

22 *IEEE Standard for Information Technology–Telecommunications and Information Exchange between Systems - Local and Metropolitan Area Networks–Specific Requirements - Part 11: Wireless LAN Medium Access Control (MAC) and Physical Layer (PHY) Specifications*, in IEEE Std 802.11-2020 (Revision of IEEE Std 802.11-2016), vol., no., pp.1–4379, 26 Feb. 2021, doi: 10.1109/IEEESTD.2021.9363693.

23 Part 11: Wireless LAN Medium Access Control (MAC) and Physical Layer (PHY) Specifications (1997). IEEE 802.11-1997, IEEE,.

24 Selinis, I., Katsaros, K., Allayioti, M. et al. (2018). The race to 5G Era; LTE and Wi-Fi. *IEEE Access* **6**.

25 IEEE 802.11b-1999(1999). Part 11: Wireless LAN Medium Access Control (MAC) and Physical Layer (PHY) Specifications: Higher-Speed Physical Layer Extension, IEEE.

26 Flores, A.B., Guerra, R.E., Knightly, E.W. et al. (2013). IEEE 802.11 af: a standard for TV white space spectrum sharing. *IEEE Communications Magazine* **51** (10): 92–100.

27 Kang, H., Lee, D. and Jeong, B.-J. (2011). Coexistence between 802.22 and 802.11af over TV white space. *International Conference on ICT Convergence*, pp. 533–536.

28 Part 19: TV White Space Coexistence Methods. (2014). IEEE 802.19.1-2014.

29 Part 19: Wireless Network Coexistence Methods. (2018). IEEE 802.19.1-2018.

30 Filin, S., Baykas, T., Harada, H. et al. (2016). IEEE standard 802.19.1 for TV White space coexistence. *IEEE Communications Magazine* **17** (6): 22–26.

31 Part 15.2. (2003). Coexistence of Wireless Personal Area Networks with Other Wireless Devices Operating in Unlicensed Frequency Bands, IEEE.

32 Part 15.1. (2002). Wireless Medium Access Control (MAC) and Physical Layer (PHY) Specifications for Wireless Personal Area Networks (WPANs), IEEE 802.15.1-2002.

33 Ennis, G. (1998). Impact of Bluetooth on 802.11 direct sequence, IEEE P802.11-98/319.

34 Golmie, N. and Mouveaux, F. (2001). Interference in the 2.4 GHz ISM band: impact on the Bluetooth access control performance. *IEEE International Conference on Communications*, vol. 8, pp. 2540–2545.

35 Howitt, I. (2001). WLAN and WPAN coexistence in UL band. *IEEE Transactions on Vehicular Technology* **50** (4): 1114–1124.

36 Part 15.1. (2005). Wireless Medium Access Control (MAC) and Physical Layer (PHY) Specifications for Wireless Personal Area Networks (WPANs), IEEE 802.15.1-2005.

9

LTE Carrier Aggregation and Unlicensed Access

9.1   Introduction

While mobile telephony operators have traditionally relied on exclusive access to spectrum for their deployments, there are significant advantages to be gained by moving toward a more flexible model. As discussed in Chapter 2, flexibility and coexistence mechanisms that take advantage of available unlicensed spectrum have the potential to greatly increase spectrum utilization as whole. Through a well-balanced mix of regulatory oversight and technical cooperation, it should be possible for mobile operators to increase their own capacities without unfairly occupying unlicensed spectrum.

Since its initial deployment in 2009, the UMTS long-term evolution, or simply "LTE" as it is now commonly referred to, has become the world's dominant mobile telecommunications system. According to the Ericsson Mobility Report, as of November 2018, LTE represents 45% of all global cellular connections [1]. This is illustrated in Figure 9.1, which shows North America having the largest market share at roughly 88%.

LTE market share is likely to increase in the near future due to a couple of factors. First and foremost, global mobile data traffic is likely to continue to increase. Figure 9.2 shows an example of mobile data traffic measurements between 2013 and 2018, indicating a growth rate of at least 50% per year [1]. In terms of exabytes (10^{18}bytes) exchanged between mobile networks and end user devices per month, this represents an increase of roughly 1500%.

So what is driving this continued increase in mobile data traffic? The answer is complicated and includes a number of related factors. The first factor is that end user devices are now widely available and span a large range of form factors including phones, tablets, vehicles, and more. For example, since its introduction in 2007, Apple has sold approximately 1.5 billion iPhone devices worldwide [2].

Wireless Coexistence: Standards, Challenges, and Intelligent Solutions, First Edition.
Daniel Chew, Andrew L. Adams, and Jason Uher.
© 2021 The Institute of Electrical and Electronics Engineers, Inc.
Published 2021 by John Wiley & Sons, Inc.

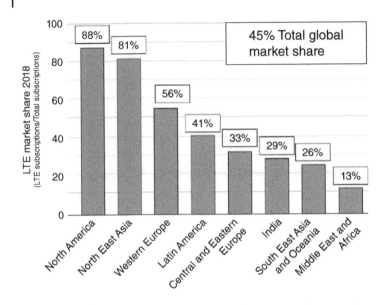

Figure 9.1 LTE market share. *Source*: Adapted from [1].

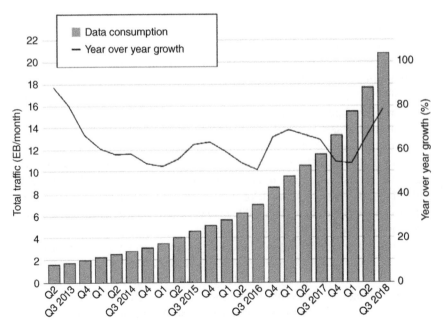

Figure 9.2 Measured voice and data traffic from 2013 to 2018. *Source*: Adapted from [1].

Since then, industry wide growth has resulted in significant competition for device market share, and although still arguably the most well-known brand, Apple smartphone sales were exceeded by Samsung in 2018, and followed closely by Huawei [3]. All three manufactures combined sold 703.4 million smartphones in 2018 alone.

The second factor making the LTE market likely to increase is the increasing amount of mobile content available. Government, private sector, and universities all seem to be taking advantage of mobile productivity applications; however, mobile data traffic is now clearly dominated by social media. In fact, according to Apple, the top 5 iOS applications downloaded in 2018 were all social media applications: YouTube, Instagram, Snapchat, Messenger, and Facebook [4].

This overall increase in mobile data traffic has shifted the focus of mobile telephony standards away from providing an increasing number of voice calls toward carrying as much IP traffic as possible. While earlier 3G deployments had the ability to carry data traffic, it was not the primary goal taken into consideration when optimizing the protocol. Beginning with fourth-generation (4G) protocols, the data carrying capacity is now the focus, dramatically changing the constraints under which the networks must operate. This chapter will first cover the paradigm shift from voice to data traffic in LTE. The methods used to shift networks from voice to data provide the necessary context for the remainder of the chapter, which outlines the problems LTE faces operating in a shared spectrum and how they are solved.

9.2 3G to LTE

Long-term evolution (LTE) is the successor to the Universal Mobile Telecommunications System (UMTS), which has been the world's dominant third-generation (3G) mobile telecommunications system since 2000. Its specifications are maintained by the third-generation partnership project (3GPP), which is a collaboration of standards bodies and contributors aimed at providing a common, stable platform for the standardization of cellular telecommunications network technologies [5].

In 2004, 3GPP began studying the "*long term evolution*" of UMTS, hence the term LTE. The goal was to develop a specification to provide increased traffic rates with low latency 10 years out and beyond [6]. The result of the study items was the **evolved packet system** (EPS), which contained two major components. The first is the **system architecture evolution** (SAE), which covered the core network. The second is LTE, which covered the **radio access network** (RAN), open air interface, and mobility. As is now evident, the entire system is commonly referred to as LTE.

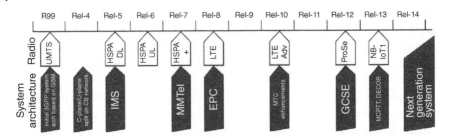

Figure 9.3 3GPP major releases. *Source*: adapted from [5].

One significant outcome of the study was the elimination of the circuit switch domain common to both UMTS, and the Global System for Mobile communications (GSM) before it. From its inception, LTE was optimized for packet data, and as such, voice is routed by way of Voice over Internet Protocol (VoIP) or by falling back to UMTS or GSM.

Figure 9.3 gives a summary of the major 3GPP releases since release 99 in 2000, which was the initial specification of UMTS. The user can follow the history of the releases using this figure. Both EPC and LTE were specified in release 8 in 2008; commercial deployments began in 2009.

It is also worth mentioning that LTE was not initially synonymous with 4G. The term "4G" was actually used by the International Telecommunications Union (ITU) to describe their separately proposed UMTS evolution, also released in 2008. The 4G requirements actually exceeded the release 8 LTE specifications in terms of peak traffic rates. However, given LTE's performance gains over UMTS, the commercial marketing community thought it best to also highlight the stark technology transition from the UMTS system architecture to EPC/LTE. As such, LTE was marketed as a fourth-generation cellular technology or 4G. The ITU agreed in 2010, making LTE synonymous with 4G ever since.

9.3 LTE Coexistence Strategies

While LTE and the other 3GPP standards are clearly designed to be deployed in scenarios where the operator has full control over the spectrum, there has been an increasing push to adopt some basic coexistence methods to take advantage of all possible spectrum opportunities. In addition, support for roaming between different operator networks can be thought of as "coexistence at the network level" and the strategies used for handover are something that could easily be applied to future RAT concepts. There are four primary technologies added to the 3GPP specifications that support spectrum flexibility for a mobile operator: small cells, dual connectivity, carrier aggregation, and licensed-assisted access (LAA). The first three in this list are specifically targeted at allowing operators to be more flexible in the way that they deploy networks. For all three, the operator still holds an exclusive right to that block of spectrum but the frequency agility of these features allows for a sort of "self-coexistence" that ensures spectrum is being used efficiently as possible. The fourth technology, LTE licensed-assisted access (LTE-LAA), is the only one that has specific allowances for coexistence strategies but it is also heavily enabled by the strategies of other three. Here, uplink and downlink traffic are routed through unlicensed spectrum while control and scheduling are still managed over the traditional licensed carrier. Given that all traffic in unlicensed spectrum must compete with others for channel access under the rules set by the regulatory bodies, LTE-LAA is especially relevant to our discussions. We explore its design goals and implementation for the rest of this chapter.

9.4 LAA Motivation

The underlying reasons and socioeconomic impact of social media popularity are topics for another text; our interest is the adaptation of LTE to the increased mobile data traffic discussed above, including the push into underused and unlicensed spectrum bands. The methods by which LTE supports an increased number of end users, an increase in individual traffic rates, and maintains quality of service (QoS), all with finite spectrum allocations are germane to our exploration of wireless coexistence strategies.

As suggested by Figure 9.2, mobile traffic rates are likely to increase for the foreseeable future. As such, it is informative to review the mathematical expression for channel capacity as described by the Shannon-Hartley theorem shown in Eq. (9.1).

$$C = B * \log_2\left(1 + \frac{S}{N}\right) \tag{9.1}$$

Channel capacity C in *bits/Hz*, is proportional both to the bandwidth B in *Hz*, and to the binary logarithm of the signal to noise ratio S/N. This is to say that if we want to increase the amount of data sent over a particular channel in a specific amount of time, we have two choices. The first is to increase the symbol rate at the expense of bandwidth consumption; the second is to increase the constellation size at the expense of required signal-to-noise ratio for correct symbol detection. The optimization of these two parameters is typically driven by system constraints including limitations on existing deployments and country-specific operating regulations. For LTE-LAA, 3GPP decided to focus on increasing bandwidth.

Given its scarcity and cost, acquiring additional licensed bandwidth may not be feasible. Furthermore, the bandwidth increase need not be contiguous as traffic from multiple *component* carriers can be aggregated above the radio access network. In fact, carrier aggregation has been specified since release 10 in 2011. However, without access to additional licensed spectrum, new component carriers must be placed in unlicensed spectrum. Use of unlicensed spectrum is not entirely new; mixed 3GPP/WLAN deployments have been specified since release 6 in 2004 [7]. Users can access the core network via WLAN access points rather than a 3GPP RAN. This is especially useful in geographical areas were WLAN connectivity far exceeds that of a 3GPP RAN. LTE-LAA does not mimic this same deployment model; it does however show how LTE can be enhanced with other commercial technologies.

3GPP began a feasibility study into the direct use of unlicensed spectrum in 2015. Among other topics, technical report (TR) 36.889 [8] reviewed spectrum considerations, deployment scenarios, design goals, and coexistence strategies. Its major findings are summarized in the following paragraphs.

The first finding of the report was that unlicensed spectrum cannot replace licensed spectrum. The licensed bands are required to ensure reliable connectivity and high quality-of-service for end users. Use of unlicensed spectrum should be viewed as complementary in an overall strategy to address increased traffic growth.

The second finding was that unlicensed access should target the 5150–5925 MHz frequency range. Exact frequencies and power levels are country-specific, but the majority is available in most countries. Furthermore, 5150–5925 represents hundreds of MHz still under relative light loading, as compared to the 900 and 2400 MHz bands. Those bands are much smaller, and already support many popular wireless technologies including Bluetooth and Wi-Fi. Lastly, with respect to user equipment (UE) hardware complexity, this band can still be covered by a single filter.

Third, the new unlicensed framework should contain adequate configurability such that the eventual deployment can operate within the various regional regulations. As mentioned above, although most countries support unlicensed access to some or all of the 5 GHz band, emissions restrictions are country-specific.

Finally, any design should support fair coexistence with both Wi-Fi and multiple service providers. Existing Wi-Fi services should not be impacted with respect to throughput and latency. Furthermore, all service providers operating within a common geographical area should be able to achieve similar performance with respect to throughput and latency.

9.5 LTE Overview

9.5.1 Evolved Packet System

The evolved packet system, or "LTE" as is now commonly known, is depicted in Figure 9.4. It contains (3) major network components: **user equipment** (UE); **radio access network** (RAN); and **evolved packet core** (EPC) [6, 9]. UE are the end user devices being serviced by the various LTE service providers. Common examples include smart phones, tablets, and personal computers. UE hosts the user applications seeking access to packet data networks (PDN), such as the Internet and private corporate networks. To service these applications, the UE communicates with the EPC via the RAN. The EPC in turn communicates with the PDN to service the user application.

The RAN communicates with the UE via the LTE air interface, and with the EPC via the *S1 interface*. The EPC communicates with the PDN via the *SG interface*. The S1 and SG interfaces are *logical* connections implemented as transport protocols on an underlying internet protocol (IP) transport network [6]. The details of the S1 and SG interfaces exceed the scope of this discussion, except as a means to illustrate the progression of data from the UE to the packet data network (Internet).

The LTE air interface, E-UTRAN, is also a transport protocol; however, here the physical layer is the physical signaling specified by technical specifications (TS) 36.211, 36.212, and 36.213, rather than over Ethernet. This physical transport protocol is further broken down into two separate logical domains: the **control plane** and the **user plane**. As their names imply, the control plane carries data used by the RAN and UEs required to handle network functions such as initial network attachment, handoffs, bandwidth requests, and call signaling. In contrast, the user plane carries the application-specific data requested by the user of each UE. The logical separation of these data planes allows for a hybrid control and data delivery model that would allow for different air interfaces to be used for the different

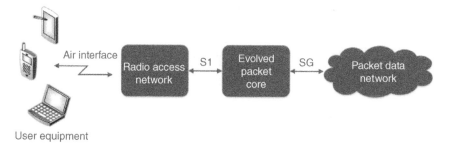

Figure 9.4 System architecture evolution.

functions. For example, the authentication, control, and monitoring of the network can be done on a very reliable but low data rate air interface while the actual data served to the user could be sent via a separate high bandwidth method.

In the next few sections, we discuss the EPC, RAN, and air interface details necessary to motivate our discussion on LTE-LAA. We reiterate that UE is a major component of an LTE network; however, since the RAN manages both uplink (UE to EPC) and downlink (EPC to UE) traffic, we fold relevant UE topics into the RAN and air interface discussions to increase clarity and limit redundancy.

9.5.2 Evolved Packet Core

The EPC communicates with PDN's to exchange the information necessary to service applications running on the UE. It is composed of multiple nodes supporting both control plane and user plane functionality. Figure 9.5 depicts the major nodes and interfaces; although a single node of each type is shown, there are typically multiple of each type depending upon the geographical area, specific implementation, and network capacity.

The control plane is composed of **mobility management engine** (MME) nodes. Each UE device is assigned to a single MME, which manages its network connection and release, active and idle state transitions, and encryption key exchange [6, 9]. This specific MME is commonly referred to as the device's *serving*

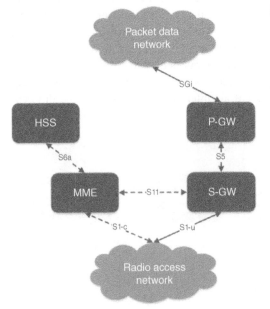

Figure 9.5 Evolved packet core.

MME. In addition to the initial serving MME, the device may be handed off to other MME nodes as the device moves across a geographical region. Each MME communicates with a home subscriber service (HSS) node, which maintains a database of the all network's subscribers.

The user plane is composed of **serving gateway** nodes (S-GW), which route data between the RAN and the EPC. Similar to an MME, each device is assigned to a single S-GW but will be handed off to other S-GW nodes as its moves through a geographical region. The **packet data network gateway** (P-GW) nodes exchange information with external packet data networks including the internet and private cooperate networks. As such, the P-GW handles IP address arbitration and quality of service (QoS) as enforcement. Each device is initially assigned to a *default* P-GW during network connection but can be assigned to multiple P-GWs as it requests access to multiple PDNs.

9.5.3 Radio Access Network

The RAN provides UE wireless access to the EPC. As such, its functions include downlink transmissions to UE, uplink transmissions from UE, transmission scheduling, radio resource management, and retransmission protocols [9]. The RAN comprises a single logical element, the **evolved node B** (eNB). The eNB is commonly implemented as a single base station supporting UE in (3) separate sectors or "cells". This scenario is illustrated in Figure 9.6.

Although there may be multiple eNB within range of each UE, it "attaches" to a single cell only. The selection process is based upon multiple factors, including cell compatibility and perceived signal strength. After initial selection, the UE

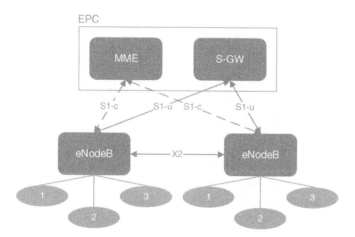

Figure 9.6 Radio access network. *Source*: Modified from Dahlman et al. [9].

periodically re-evaluates its cell selection, and moves to another cell if selection criteria dictates.

In Figure 9.6, two eNB nodes are shown communicating with each other over the *X2 interface*. The details of the X2 interface exceed the scope of this discussion. What is important is that the communication supports multicell resource management, intercell interference coordination, and UE mobility. As mentioned earlier, eNB communicate with the EPC over the S1 interface. For completeness, Figure 9.6 shows both eNB connected to single MME and S-GW instances. As was discussed earlier, the MME represents the EPC control plane and the S-GW represents the EPC user plane.

9.5.4   Air Interface

The air interface protocol architecture is illustrated in Figure 9.7 [6][9]. Starting from the bottom, the physical layer (PHY) performs all the analog and digital signal processing functions supporting transmissions between the eNB and the UE. These include modulation/demodulation, coding/decoding, and antenna mapping. The PHY provides services to the medium access control layer (MAC) in the form of transport channels. The MAC layer maps logical channels to transport channels, schedules transmissions between the UE and the eNB, and manages hybrid-ARQ retransmissions. The MAC layer provides services to the radio link control layer (RLC) in the form of logical channels. The RLC layer ensures in-sequence delivery of data to higher layers. This includes segmentation/concatenation of packet data, retransmission per the hybrid-ARQ protocol, and duplicate removal. The RLC provides services to the packet data convergence protocol layer (PDCP). The PDCP layer performs various compression and ciphering schemes, and assists the RLC in duplicate removal in case of handover between cells.

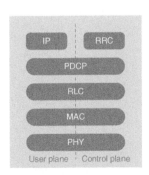

Figure 9.7 also illustrates the use of control and user planes. Protocols executed in the user plane process data that is of interest to users, e.g., IP packet data. In contrast, protocols executed in the control plane process data that is of interest to the network only; typically that used for network maintenance and resource management. For example, the radio resource control (RRC) layer manages multiple functions required for UE attachment and mobility, including the broadcast of system information, cell reselection, and device capability discovery.

Figure 9.7   Air interface protocol architecture.

Both the uplink and downlink of the LTE air interface utilize orthogonal frequency division multiplexing (OFDM) (*See Chapters 3 and 4 for a description of*

OFDM and OFDMA respectively) with a variety of configurations to ensure that the RAN can adjust the link for evolving channel conditions, user allocations, and user-scheduling schemes. The OFDM configurations for both the uplink and downlink resource blocks are dynamically allocated according to the changing channel conditions. This style of dynamic assignment allows for flexible deployment models, depending on the spectrum allocations available to any single operator [9]. For example, an operator could have a configuration that uses a traditional spectrum model with the downlink on a particular frequency and the uplink on a totally different one known as frequency-division duplexing (FDD). However, they could also use a single-frequency channel for both the uplink and downlink called time-division duplexing (TDD). TDD and FDD are discussed in detail in Chapter 4. In this case, the optimal methods used to combat rapidly changing channel conditions, such as those found in high-speed mobile users, can utilize the ideal OFDM configuration for a TDD channel. In addition, multiple transmit and receive antennas at both the eNB and UE provide spatial diversity, interference suppression, beam forming, and frequency reuse [9].

9.6 Carrier Aggregation

Carrier aggregation gives service providers the ability to utilize fragmented spectrum allocations [6]. For example, if allocated 5–10 MHz swaths of spectrum across several different frequency bands, carrier aggregation allows service provides to treat this as a single contiguous allocation. The result is increased uplink or downlink traffic rates as compared to that were carrier aggregation not supported. Initial design of this capability was heavily motivated by anticipated use of large swaths of unlicensed spectrum. The LTE-LAA model is based heavily on carrier aggregation, and therefore a strong understanding of the CA concepts is required to understand LAA.

Three unique scenarios are illustrated in Figure 9.8 [9], the first being intraband aggregation with contiguous carriers. Here traffic is sent over carriers 1 and 2 of frequency band A. The second scenario shown is intraband aggregation with non-contiguous carriers. Here traffic is sent over carriers 1 and 3 of frequency band A. The third scenario shown is interband aggregation. Here traffic is sent over carrier 1 of frequency band A and carrier 3 of frequency band B.

Each carrier is referred to as a *component* carrier and is backward compatible with releases that do not support carrier aggregation. A primary carrier is assigned to both uplink and downlink to support essential control signaling. For example, secondary carriers are deactivated when the UE is placed in idle mode and are reactivated when the UE resumes network activity [9].

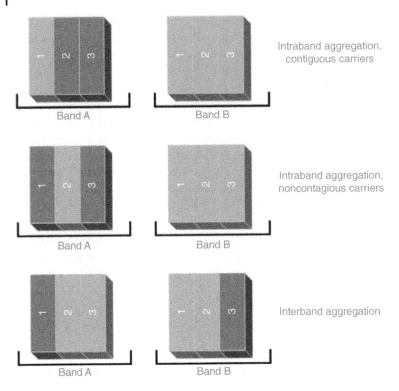

Intraband aggregation, contiguous carriers

Band A   Band B

Intraband aggregation, noncontagious carriers

Band A   Band B

Interband aggregation

Band A   Band B

Figure 9.8   Carrier aggregation.

The current aggregation limit is 32 carriers, resulting in an aggregate bandwidth up to 640 MHz. These can be a combination of both FDD and TDD carrier configurations, and all radio frame types, with the only assumption being UE capability. For example, if a UE does not support full-duplex operation, it cannot receive traffic on a secondary downlink carrier while transmitting on the primary uplink carrier. Although not mandated, carrier aggregation is typically asymmetric, with the downlink allocated more component carriers than the uplink [5][9].

Referring back to our discussion on the RAN protocol architecture and Figure 9.7, carrier aggregation duplicates the PHY and MAC layer processing for each component carrier. However, the rest of the protocol stack remains unchanged as the RLC and above treats all component carriers as a single entity. The MAC entity manages data distribution across the component carriers, and each receives its own hybrid-ARQ process.

Recall that component carriers are not limited by duplex type or frame structure. As such, both the traffic and respective acknowledgments follow an array of scheduling strategies. Initially, these break down into two main categories: self-scheduling

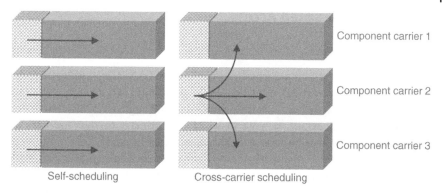

Figure 9.9 Component carrier scheduling.

and cross-scheduling. In **self-scheduling**, the control region of each component carrier specifies the traffic on that carrier. In **cross-scheduling**, the control region of a component carrier specifies traffic on one or several component carriers. Both schemes are illustrated in Figure 9.9 [9].

For both FDD and TDD self-scheduling, and FDD cross-carrier scheduling, each UE searches for its downlink traffic on all available subframes. For TDD cross-carrier scheduling, however, the UE must obtain the location of its frames from the control information sent by a scheduler.

Over time, the 3GPP has increased the capabilities provided by carrier aggregation. Starting with initial capability in Release 10, each successive release included new capabilities such as allowing different configurations between carriers in Rel11, supporting combined TDD and FDD in Rel12, and increasing the number of supported carriers in Rel13. These are summarized in Figure 9.10. In addition, future study items and working group research tasks continue to improve the efficiency and reliability of carrier aggregation to provide operators with the necessary flexibility to maximize their spectrum licenses [9].

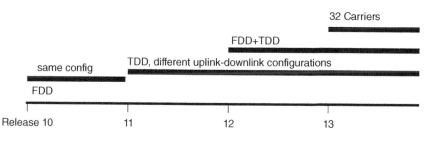

Figure 9.10

9.7 License-Assisted Access

9.7.1 Basic Concepts

We finish this chapter with a discussion on LTE-LAA. LAA is designed to increase network throughput without access to additional licensed spectrum and without sacrificing quality-of-service. This is primarily done through a modification to the existing carrier aggregation scheme in which the primary carriers are located within licensed spectrum and the secondary carriers are located within unlicensed spectrum. This is illustrated in Figure 9.11 for the case of one primary uplink/downlink pair, and one secondary carrier uplink/downlink pair.

As is the case with carrier aggregation in licensed spectrum, multiple secondary carriers can be configured for each user. These also need not be contiguous in frequency as LAA is capable of taking advantage of fragmented spectrum. Furthermore, because this is unlicensed spectrum, fragmentation must be addressed in both time and frequency [9].

For carrier aggregation, the designation of a primary carrier facilitates eNodeB maintenance of essential control signaling when the UE is idle. The same is true for LAA, which also carries additional restrictions placed on unlicensed spectrum usage around the world. Each region promotes fair use of unlicensed spectrum through regulations aimed at providing equal access to competing commercial technologies. For the 5 GHz band, these include WLAN (Wi-Fi) and multiple LTE service providers. Each region also ensures ample availability of unlicensed spectrum for critical infrastructure and national defense. These typically include municipal and federal communications and radar systems. These are given priority, such that if detected, the channel must be vacated immediately for a fixed time period. Both of these concerns result in a variety of regional operating requirements on channel usage and transmit footprint.

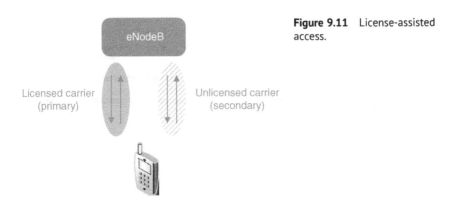

Figure 9.11 License-assisted access.

9.7.2 Deployment Scenarios

Given the usage constraints of the 5150–5925 MHz band, LTE-LAA deployment is targeted toward operator-deployed low-power nodes. Examples include dense urban areas, indoor shopping malls, and office buildings. To illustrate further, one deployment scenario suggested in the feasibility study is shown in Figure 9.12. The macro cell operates as normal over licensed spectrum; the small cell is deployed to support unlicensed spectrum only, and connects to the core network through an ideal, non-colocated backhaul to the macro cell. As an example, this could be a cluster of low-power nodes installed in an indoor shopping mall.

In operation, the UE uses LAA to aggregate its primary carrier, always located in licensed spectrum, with secondary carriers that may be located in any combination of licensed and unlicensed spectrum. Control signaling is sent over the primary carrier, and multiple secondary carriers are used to increase downlink traffic rates for high bandwidth applications. Without LAA, the increased traffic rate would come at the expense of other users' throughput, which may degrade their quality-of-service.

Additional deployment scenarios described in the feasibility study include aggregation across multiple small cells, where the secondary carriers are a mix of both licensed and unlicensed spectrum. Ideal, colocated backhaul to the licensed small cell is assumed, which removes this requirement from any macro cell also within range. If, however, ideal backhaul to the macro cell is also available, then aggregation can include secondary carriers from the macro cell as well.

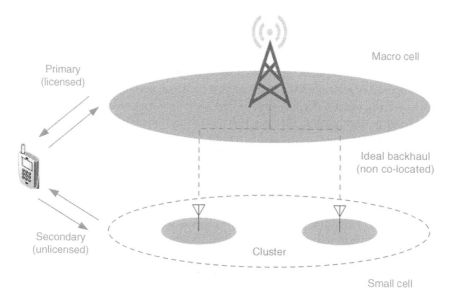

Figure 9.12 LAA deployment scenario.

9.7.3 LAA Coexistence Mechanisms

Prior to LAA, 3GPP radio access technologies relied heavily on the assumption of exclusive channel rights. First, the entire data framing structure used by LTE for both the uplink and downlink requires that users be able to transmit and listen at predetermined times. Under the current model, UEs can only know information about the data being sent if they previously obtained that information from frames sent at fixed intervals. Also, feedback about the channel state and requests for retransmissions have strict timing requirements to allow the eNodeB the time it needs to properly schedule future time/frequency resource blocks. Finally, because LTE signals are *constant carrier*, the complexity of receiver implementations is greatly reduced as timing synchronization at all levels can be tracked very precisely by following the transmitter. With rules and restrictions placed on the use of unlicensed bands in many areas, most notably restrictions on how long someone can transmit, the LTE mechanisms which rely on strict timing become invalid [9]. LAA essentially needs to compensate for the lack of ability to transmit continuously. This primarily centers on adapting the frame, subframe, and slot concepts such that the need to transmit and receive at predetermined times is removed. Also, LAA must implement specific mechanisms for channel access. These include searching for idle channels, and avoiding those currently occupied. We can describe these as three distinct coexistence mechanisms; discontinuous channel access, dynamic frequency selection, and clear channel assessment.

From a network level, and even to a certain degree the RAT level, LAA is treated no differently than any other secondary carrier in a carrier aggregation scheme. This design choice has a number of advantages and disadvantages when it comes to unlicensed spectrum usage. On the one hand, higher layers can operate in exactly the same way as they would regardless of the spectrum type being used. This ensures that operators can be as flexible as a possible without being drowned in configuration management, introducing a high load on the command and control interfaces, or even needing to define an additional control interface. The benefit here is that, from the perspective of the network, LAA could be indistinguishable from a licensed secondary carrier. The disadvantage of this approach lies in the fact that the design decisions that drove the original design for carrier aggregation relied on the assumption of exclusive channel use. By mostly sticking with this original design, designers must work around the bursty and uncertain nature of unlicensed bands.

Overall, there are three primary concepts that allow traditional carrier aggregation to transition from licensed to unlicensed bands. The first, and arguably most important, is the idea of discontinuous channel access whereby the eNB is prohibited from constantly transmitting on a single unlicensed channel, sometimes required by law but in general required just to be a good steward of the commons. The next two technologies focus on how to access the channel. **Clear channel**

assessment (CCA) tells the operator when a channel is busy and should not be interfered with and **dynamic frequency selection** (DFS) allows the operator to be agile in where it is transmitting to minimize interference.

9.7.3.1 Discontinuous Channel Access

One of the primary assumptions that drives decision making in the LTE specifications is the idea of continuous channel access, i.e. that an operator will have exclusive access to the spectrum and be allowed to transmit at any point in time at will. When it comes to unlicensed spectrum bands, this assumption is no longer valid as everyone is sharing the same spectrum and there are no guarantees about when the eNB or UE may be able to transmit.

Discontinuous channel access to secondary carriers presents two new challenges, such as resource block scheduling and synchronization. The first challenge arises from the fact that UEs can no longer rely on predetermined times to listen for downlink traffic, listen for uplink traffic allocations, or send uplink traffic due to the uncertainty of channel availability. The second challenge, UE synchronization, arises from the fact that the eNodeB synchronization signals are no longer persistent. As such, again depending upon channel availability, any prior UE synchronization may quickly become invalid and requires the UE to resynchronize before trying to communicate with the eNodeB [9].

Resource block scheduling is handled by relaxing the requirements on the traditional frame and subframe boundaries. If the traditional strict timings were enforced, the potential number of channel accesses by the secondary carrier would be very low since it would be restricted to the subset of subframe start times *and* the availability of the channel. This issue is illustrated in Figure 9.13 where the initial slot scheduling depicted in (a) is not possible due to channel occupancy from other users (b). If the strict timings were adhered to, the LAA network would not be able to transmit even though the channel is, relatively, unoccupied because the other users are occupying the channel at the start of frames.

To increase the probability of taking control of the channel, LAA attempts to establish subframes on traditional timing epochs; if unsuccessful, partially filled subframes are used that can start at any point in time as shown in Figure 9.13d. If there is sufficient data to be sent, *only* the first frame will be a non-traditional subframe. Because the transmitter has control of the channel, subsequent subframes will resume normal operation until the transmission has ended.

It is interesting to note that, although unlicensed spectrum access models are traditionally much closer to the TDD model in LTE, the LAA frame structure is nearly identical to the FDD structure. This arises from the fact that, while TDD is generally more suitable for single-frequency use, LAA is much closer to FDD because the direction of transmission means that only one user, the eNB, will be accessing the channel at any given time. Because of the similarity to the

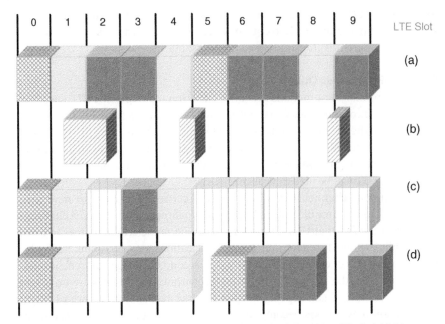

Figure 9.13 Strict LTE scheduling vs Relaxed LAA scheduling (simplified). (a) LAA transmissions as scheduled. (b) Other users. (c) Strict LAA transmissions as realized. (d) Relaxed LAA transmissions as realized.

FDD frame structure, there are a number of signals sent by the eNB that would be expected to arrive at certain times. Because, as discussed previously, there are no guarantees that the channel will be available at that time, LAA further relaxes the requirements on the eNB by allowing these critical control messages to show up in any frame. This allows the scheduler to ensure that periodic control messages are sent at the necessary intervals for operation without requiring that they be in a predefined subframe.

UE synchronization is solved through a combination of mechanisms. First, some types of LTE messages already include fields designed to help receivers synchronize with the transmitter. In traditional LTE deployments, these are used to ensure that the eNB and UE can track the changing propagation environment over time and adjust their transmissions accordingly. In LAA, however, these same signals can be used by a receiver to get an accurate assessment of the timing offsets and properly synchronize to those frames. When these frames are not present, however, the receivers still need some way to periodically synchronize with the transmitter. In this case, LAA employs a technique previously used in the LTE specification to support dynamic carriers: the **discovery reference signal** (DRS). The DRS was created out of a need for allowing UEs to stay synchronized with carriers that are frequently turned on and off to support dynamic network data loads [9]. The DRS is a composite of most of the synchronization signals sent by an eNB, but encapsulated into a single transmission rather than having the signals spread across several subframes. Utilizing the DRS for LAA allows devices to stay synchronized to the carrier provided that the messages are sent on sufficient minimum intervals. The key point here is that the signals are not required to be sent at a specific time, just at a sufficient rate.

By treating secondary carriers located in unlicensed bands as traditional carrier aggregation secondary carriers, the issue of discontinuous channel access had the potential to break a number of the scheduling and timing mechanisms inherent in the carrier aggregation mechanisms. However, by making small tweaks to those mechanisms and generally relaxing timing requirements LAA can overcome those issues.

9.7.3.2 Dynamic Frequency Selection

When accessing unlicensed spectrum, channel availability cannot be guaranteed a priori. For this reason, LAA implements a process called **dynamic frequency selection** (DFS) for selecting where in the spectrum to place an LAA subcarrier. DFS is designed to find the best channel for each secondary component carrier based upon two primary goals: maximum channel utilization and minimize interference with other users [9]. For this reason, DFS is a continuous process whereby the eNB characterizes the status each channel to decide its suitability to support traffic. As with all unlicensed access, the number one goal is to prevent interference with licensed users in the same or adjacent bands. For this reason, the first thing an LAA eNB does is check for the presence of incumbent users and vacate the channel if discovered. In addition to avoiding incumbents, the DFS mechanisms also play a role in avoiding other unlicensed users. If two unlicensed users are currently sharing a channel when there is an unoccupied channel available, it is clearly more efficient for one of those users to move, providing a full channel to each user.

While the LAA standards do not call out a specific method or algorithm for performing DFS, it is clear that these algorithms are a critical component to LAA operation. It is accepted that, at a minimum, specific implementations will be required to comply with regional regulations surrounding channel access. For example, the amount of time that a secondary user must vacate a channel after identifying an incumbent varies across regulatory zones; each LAA deployment will need to be configured to match local regulations to provide the most efficient use of bandwidth.

DFS does come with drawbacks, however. Moving to a new channel can be costly in terms of overhead and user management, so ideally this is an action that would be taken infrequently. If maximizing use of the available spectrum is the ultimate goal for LAA, the DFS mechanisms can be thought of as a "macro"-level decision that is made on a larger time scale compared other methods of short-term spectrum sharing such as the CCA discussed in the next section. Overall, the DFS mechanism provides carriers with the flexibility to move to unoccupied spectrum rather than applying real-time coexistence mechanisms that reduce the overall efficiency of everyone using the channel.

9.7.3.3 Clear Channel Assessment

As mentioned previously, the primary challenge that makes LAA different from traditional LTE carrier aggregation is the uncertainty of channel availability. The process of determining availability is called **clear channel assessment** (CCA). For the bands currently targeted by LAA, its primary competitor for channel access will most likely be the 802.11 family of standards, also called Wi-Fi. For this reason, 3GPP has tailored the LAA CCA mechanisms to match closely that of the 802.11; this is done by employing a similar **listen before talk** (LBT) scheme [9]. Each UE must monitor a channel for a certain period of time to verify that no one else is utilizing that channel. While this scheme is simple on its face, there is a significant amount of configuration and customization that can be done to enable access priorities and minimize collisions between different transmitters.

The LBT scheme can be described simply by the following: when a node wants to transmit data, the channel in question must be clear for a minimum amount of time before that node can begin transmitting. This value depends upon a number of factors including the operating band, the desired individual throughput, and the probability of fair access to all radio access technologies sharing the spectrum. Here we refer to this value as the *total backoff time*.

First, it is important to note that the total backoff time need not be contiguous. In fact, when executing LBT, an LAA node must cease counting time when someone else is transmitting. However, it does not have to reset its timer, which would result in unlucky nodes infrequently gaining channel access in busy networks. Second, calculation of the total backoff time is dependent on the class of intended traffic. The portion of time that different classes must wait is referred to as the *defer period*. A smaller defer period reduces the total backoff time for more important traffic and greatly improves the chance of taking control of the channel. For example, signaling and real-time data are given a high priority, and therefore, utilize a lower defer period. Finally, the total backoff time for LAA always includes a random component referred to as the *random backoff time*. This random component is added to ensure that all nodes with pending transmissions do not try to transmit simultaneously at the first free opportunity, as this would essentially guarantee collisions in even lightly loaded networks. For both LAA and Wi-Fi, this time is chosen using a random number referred to as the *contention window*, multiplied by the backoff slot time as shown in Eq. (9.2).

Table 9.1 LBT durations for different traffic classes in IEEE 802.11 [10] and LTE LAA downlink [11].

Traffic class	Minimum wait time		Contention window minimum		Contention window maximum	
	802.11 (AIFS)	LAA (Defer)	802.11	LAA	802.11	LAA
1	25us	25us	3	3	7	7
2	25us	25us	7	7	15	15
3	43us	43us	15	15	63	63
4	79us	79us	15	15	1023	1023

Sources: IEEE 802.11-2016 [10] and ETSI [11].

$$\text{Total Backoff} = \text{Defer Period} + \text{Contention Window(Backoff Slot)} \quad (9.2)$$

To summarize, to utilize a potentially available channel, a radio must always wait the minimum defer period based on the data class and then wait an additional random amount of time for the channel to be clear. Table 9.1 shows the standard values for these numbers in both 802.11 and LTE LAA.

Looking closely at Table 9.1, it is clear that one of the goals during the design phase of LAA was to remain as compatible as possible with 802.11 when sharing unlicensed spectrum. By selecting a similar CCA procedure and matching contention periods, fair use can be assured in the sense that LAA carriers and 802.11 networks will have equal priority for accessing the channel [12].

9.8 Deployment Status

Since its initial release, LTE LAA has been supported by both carriers and hardware manufacturers. In their 2019 report on unlicensed spectrum access, the Global mobile Suppliers Association (GSA) determined that there were 37 operators investing in LAA deployments spread across 21 different countries [13]. Of the operators investing in LAA, 20 are in the field trial stage and nine have operationally deployed systems. Of course, deployment of these systems is not worthwhile unless there are customers to use them. When looking specifically at user equipment, the GSA found that there were 90 consumer devices available supporting LAA in a variety of form factors including mobile phones, mobile hotspots, industrial modules, and tablets. Though it remains to be seen if LAA will introduce significant interference in practice, it is very likely that LAA as a technology is going to see widespread adoption wherever the designated 5 GHz spectrum allocation is underutilized.

9.9 Conclusions

Mobile telephony and data services continue to be one of the largest consumers of wireless spectrum resources with no apparent end in sight. As the cost of spectrum grows, by necessity, those costs will be passed on to consumers. It is right, then, to ask ourselves if the use of unlicensed spectrum by private telecommunications companies is better in the long run for the public good. It is clear that allowing access to this unlicensed spectrum provides for an overall increase in bandwidth available to end users, and at a lower operating cost. Relying on free market principles, it then stands to reason that this would provide an overall economic benefit to consumers, including higher data rates, lower overall service costs, or both. However, these benefits must be balanced against the idea of a *free and open* use model for unlicensed spectrum whereby commercial carriers are not allowed to monopolize a public resource.

While there is no clear mandate either way, the availability of enabling technologies such as LTE LAA allows governments, industry, and regulatory bodies to explore the different possibilities in a limited way, providing real-world results that can be used to make broader regulatory decisions about the use of unlicensed spectrum by commercial entities.

References

1 Ericsson. (2019). Ericsson Mobility Report Q4 2018 Update. https://www.ericsson.com/en/mobility-report/reports/q4-update-2018.

2 Statista. (2019). Unit sales of the Apple iPhone worldwide from 2007 to 2018 (in millions).https://www.statista.com/statistics/276306/global-apple-iphone-sales-since-fiscal-year-2007.

3 Sui, L. (2019). Global Smartphone Shipments Declined on a Full-Year Basis for First Time Ever in 2018.https://www.strategyanalytics.com/strategy-analytics/blogs/devices/smartphones/smart-phones/2019/01/31/global-smartphone-shipments-declined-on-a-full-year-basis-for-first-time-ever-in-2018.

4 Yurieff, K. (2018). The most downloaded iOS apps of 2018.https://www.cnn.com/2018/12/04/tech/ios-most-popular-apps/index.html.

5 3GPP. (2019). 3GPP.https://www.3gpp.org/about-3gpp/about-3gpp.

6 Cox, C. (2014). *An Introduction to LTE*. West Sussex: Wiley.

7 Forsberg, D., Horn, G., Moeller, W.-D., and Niemi, V. (2013). *LTE Security*. West Sussex: Wiley.

8 Mazzarese, D. (2015). TR36.889 R13.https://www.3gpp.org/ftp/Specs/archive/36_series/36.889.

9 Dahlman, E., Parkvall, S., and Skold, J. (2016). *4G LTE-Advanced Pro and the Road to 5G*. London: Academic Press.

10 IEEE Std 802.11-2016 (2016). *IEEE Standard for Information technology—Telecommunications and information exchange between systems Local and metropolitan area networks—Specific requirements - Part 11: Wireless LAN Medium Access Control (MAC) and Physical Layer (PHY) Specifications*. IEEE Std 802.11-2016 (Revision of IEEE Std 802.11-2012), pp. 1–3534, doi: https://doi.org/10.1109/IEEESTD.2016.7786995.

11 ETSI. (2017). LTE; Evolved Universal Terrestrial Radio Access (E-UTRA),*ETSI TS 136 213, V14.2.0*.

12 Mehrnoush, M., Roy, S., Sathya, V., and Ghosh, M. (2018). On the fairness of Wi-Fi and LTE-LAA coexistence. *IEEE Transactions on Cognitive Communications and Networking* **4** (4): 735–748.

13 Global mobile Suppliers Association (GSA) (2019). LTE in Unlicensed Spectrum Report: July 2019. Global mobile Suppliers Association (GSA). https://gsacom.com/paper/unlicensed-shared-spectrum-report-july-2019/.

10

Conclusion and Future Trends

10.1 Summary of the Preceding Chapters

One of the primary goals of this book is to elucidate current wireless coexistence standards. In order to accomplish that goal, Chapters 2 through 6 provided background information relevant to understanding the growing complexity of wireless coexistence. With this information in hand, Chapters 7 through 9 are able to effectively describe published coexistence standards and the reasoning behind the decisions made at each stage of the standardization process.

Figure 10.1 illustrates the technology circle from reference [1] first introduced in Chapter 1. The first portion of this book followed along with the layered model from the diagram, beginning with regulatory frameworks in Chapter 2 and building an understanding of the Physical and Medium Access Control layers in Chapters 3 and 4. Chapter 2 detailed the evolution of spectrum regulations over time. Spectrum regulation began with a simple strategy of separating emitters in space and frequency in order to provide coexistence. These regulations have adapted and continue to adapt to increased demand over time. Chapter 3 then provided the reader a series of primers into relevant communications theory including noise, multipath effects, orthogonal frequency division multiplexing, and modern transceiver hardware. These primers provide the requisite background material to discuss the concepts in Chapter 4, which deals with the resolution of contention among wireless users with equal priority. Among these techniques is Carrier Sense Multiple Access (CSMA) used in the IEEE 802.11 Wireless Local Access Network (WLAN) standard. CSMA uses a listen-before-talk approach to mitigating contention. Another group of users required to listen-before talk are secondary users. These secondary users, users of lower priority to access the channel, must be able to sense the presence of incumbent signals before they are allowed to transmit. Chapter 5 details spectrum sensing techniques used both by primary users, as in CSMA, and by lower priority users to avoid interfering with an incumbent.

Wireless Coexistence: Standards, Challenges, and Intelligent Solutions, First Edition.
Daniel Chew, Andrew L. Adams, and Jason Uher.
© 2021 The Institute of Electrical and Electronics Engineers, Inc.
Published 2021 by John Wiley & Sons, Inc.

The processing of spectrum sensing data is becoming more complex and, thus, Chapter 6 details the concept and function of *intelligent radios* and the methods by which they may avoid interference and dynamically access spectrum. At this point, the relevant issues that must be tackled to adequately understand different coexistence protocols should be well understood by the reader.

With the knowledge of the difficulties surrounding effective wireless coexistence. Chapters 7 through 9 detailed selected wireless coexistence standards enabled by the concepts developed in Chapters 2 through 6. Chapter 7 provided an overview of the IEEE 1900 series of standards. This standard provides great insight into the field of wireless coexistence by presenting a unified ontology for wireless spectrum sharing that is used to compare the capabilities of different radio systems. Chapter 8 explored the wireless standards from the IEEE 802 series, which provided significant insight into operation in the ISM band, the traditional home for unlicensed access, and how those same concepts can be modified for application to future regulated bands specifically intended for dynamic spectrum access radios. Finally, Chapter 9 covered the largest commercial use of unlicensed spectrum, the 3GPP LTE License Assisted Access standard that uses traditional cellular telephony networks to coordinate data transfer in the unlicensed bands while maintaining compliance with the regulations on interference and transmit power in those bands.

Another important aspect to Figure 10.1 is that it shows the relationship between user demand (layer 8) and regulatory framework (layer 0). As people use more bandwidth in their daily lives, the regulations that govern the shared spectrum must provide means for industry to meet that demand. Chapter 2 details the efforts of regulatory authorities to open the spectrum to more efficient use.

One important development in the history of spectrum regulation discussed in Chapter 2 was the advent of internationally recognized unlicensed bands. These unlicensed bands gave rise to mitigation strategies in commercial wireless links in which wireless systems simply expected to be interference and sought to mitigate that interference by brute force. This strategy employed spread spectrum techniques described in Chapter 4. Despite employing spread spectrum techniques, interference proved to be a major hindrance as the use of Wireless Local Area Network (WLAN) and Wireless Personal Area Network (WPAN) devices became more popular. Coexistence strategies were examined and a monitoring strategy was then adopted in which the frequency hopping WPANs would monitor their link and remove troublesome frequency channels from their hop sets. This early exploration into unlicensed band coexistence strategies is detailed in the discussion of IEEE 802.15.2 in Chapter 8.

Another important milestone in the evolution of spectrum regulations discussed in Chapter 2 was the opening of the TV Band to secondary users. It was found that the TV Band was underutilized, and therefore opened to secondary users. These

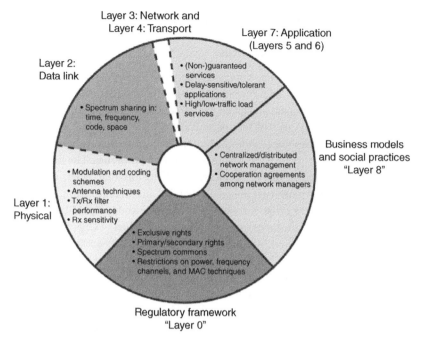

Figure 10.1 The technology circle of wireless coexistence. *Source*: Voicu et al. [1].

secondary users were required to proactively avoid interfering with the incumbent broadcasters in that band. To this end, the new wireless standards employed sensing strategies in order to determine if a frequency channel was available for use. This required the sensing techniques discussed in Chapter 5. Among the coexistence standards to deploy in the TV Bands are IEEE 802.22 and 802.11af, both discussed in Chapter 8.

The most recent development in the evolution of spectrum regulation is the establishment of the 3.5 GHz Citizens Broadband Radio Service (CBRS) band. As discussed in Chapter 2, this band frees up a large swath of underutilized spectrum that was originally reserved for important potential users. The CBRS band introduces multiple tiers of users, placing the incumbents for whom the band was originally reserved at the top and giving them priority access, and providing the lowest tier of priority to general access users.

Each chapter provides a list of references so that an interested reader can do more research on any given topic. Many of those references are recently published academic papers showing that wireless coexistence is an active area of research. So far, this book has only captured the current state of wireless coexistence techniques and associated standards. We will now turn our attention to future trends in wireless coexistence.

10.2   Nonorthogonal Multiple Access and Underlaying

10.2.1   Nonorthogonal Multiple Access

That two emitters could share the same time-frequency resource was first introduced in Chapter 4 in the Code Division Multiple Access (CDMA) discussion. In CDMA, the received signal at any given time-frequency resource is a composite of multiple transmitted signals. The idea behind CDMA is that multiple signals can share a time-frequency resource and still be separable if those signals are each spread with a unique code from a set of code with certain properties. The first property needed is *low autocorrelation sidelobes*, which means that when any one unique code is correlated with itself the result is an impulse-like function. The second property needed is low cross-correlation. Ideally the codes would be orthogonal to one another. The concept of orthogonality was detailed in Chapter 3 in the Orthogonal Frequency Division Multiplexing section. The cross-correlation product of two orthogonal signals is zero. Unfortunately, the codes used by CDMA are often not orthogonal, and the cross-correlation between different codes results in cross-correlation noise. The power of this noise is proportional to the power of the signal at the receiver. This leads to an issue called *Multi-User Interference* (MUI). In some literature, this type of interference is called Multi-user Access Interference (MAI). For CDMA systems, this type of interference leads to the *Near-Far* problem. This is where one signal contributes so much power at the receiver compared to a lower-power signal that the cross-correlation noise makes it nearly impossible to recover the lower-power signal. One resolution to this is dynamic power control, where the base station commands the emitter with the higher power to decrease its transmit power. This results in a lower cross-correlation noise allowing the receiver to recover weaker signals. Alternatively, the receiver can employ *Successive Interference Cancellation* (SIC). In a SIC receiver, the higher power signal is first demodulated then a copy of the higher power received waveform is generated and subtracted from the composite signal. Once the higher power signal is removed from the composite signal by way of this cancelation technique, the receiver can go on to process other component signals in the composite signal. From this description, we can see that CDMA is usually a nonorthogonal multiplexing scheme. Unlike other channelization schemes in which signals may ride on evenly spaced and orthogonal carrier frequencies, all user signals are bunched together in CDMA and received at once. CDMA relies upon the spreading codes to separate and channelize the users.

CDMA is but one of many Nonorthogonal Multiple Access (NOMA) schemes. Those schemes can be divided, broadly, into two families: Code-Domain and Power-Domain. CDMA belongs to the Code-Domain family of NOMA schemes. The operation of the Code-Domain schemes follows that of CDMA.

The Power-Domain NOMA schemes are different in that the signals will be deliberately separated by levels of received power. Power-Domain NOMA schemes require an interference cancelation capability at the receiver in order to separate the users.

NOMA represents a multiple access scheme that accepts interference as part of the channel analysis. Unlike other multiple access schemes that attempt to separate different emitters in frequency, time, and/or location, NOMA requires the receiver to be able to remove unwanted signals. Figure 10.2, modified from reference [2] illustrates the concept of NOMA with successive interference cancellation in the downlink of a cellular system. The base station transmits a composite signal to two mobile devices. The composite signal is composed of two unique component signals, one for each mobile device. The power transmitted by the base station is displayed next to the icon labeled "eNB" (eNodeB). The signal for the mobile device UE2 is allocated more transmit power than the signal for mobile device UE1. As discussed in Chapter 3, each mobile device experiences a different wireless channel. The mobile device UE2 demodulates the more powerful signal. The process reveals that this first signal was intended for UE2, so UE2 stops further processing as it now has its intended bits. UE1 follows the same process. The demodulation process first reveals that the more powerful signal was not intended for UE1. Therefore, UE1 creates a 180° shifted version of the component signal intended for UE2. This 180° shifted waveform is added to the composite waveform

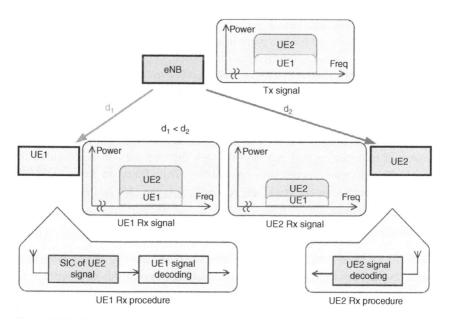

Figure 10.2 Power-domain NOMA. *Source*: Xiong et al. [2].

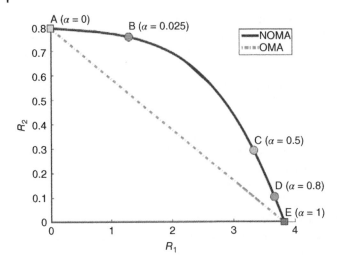

Figure 10.3 User data rates comparison. *Source*: Vaezi et al. [3].

causing destructive interference with the UE2 waveform. With the UE2 waveform removed, UE1 is now free to begin demodulating the next highest-power signal, which happens to be the one intended for UE1.

The benefit of NOMA is that the two users can share the channel capacity and enjoy a higher overall data rate. Figure 10.3 from [3] shows a comparison between the achievable data rates for two users in an Orthogonal Multiple Access Scheme and a NOMA scheme using interference cancelation. The alpha term is the scale of resource usage between the two users. If alpha is 0, then all resources go to user 2. If alpha is 1, then all resources go to user 1. For all other alphas, the two users are concurrently transmitted. The two axes in Figure 10.3 represent achievable rates for the two users. The rate is constrained by the capacity of the two wireless channels. Channel capacity is a feature of a physical wireless channel that limits the amount of information that can pass through the channel and is measured in bits/Hz. The channel capacity over an Additive White Gaussian Noise (AWGN) channel is expressed in Eq. (10.1) where $\gamma = \frac{S}{N}$, $\frac{S}{N}$ is the SNR and B is the bandwidth of the channel.

$$C(\gamma) = B \log_2(1 + \gamma) \tag{10.1}$$

For the OMA system, the possible rates over an Additive White Gaussian Noise channel are expressed in Eqs. (10.2) and (10.3) α in this case represents the share of spectrum resources, such as time.

That is to say that in OMA systems, the base station must stop using spectral resources to transmit the signal for user 1 if it is going to transmit the signal for user 2. γ_i represents the SNR for a given user (i) and is defined in Eq. (10.4) where h_i represents the channel loss for a given user (i), and P is the transmit power of the base station.

$$R_1 = \alpha C(\gamma_1) \tag{10.2}$$

$$R_2 = (1 - \alpha)C(\gamma_2) \tag{10.3}$$

$$\gamma_i = |h_i|^2 P \tag{10.4}$$

For a NOMA system, the achievable rates are expressed in Eqs. (10.5) and (10.6). In the case of NOMA, α represents the share of the transmit power of the base station dedicated to each of the two user signals. The NOMA scenario is illustrated in Figure 10.4. In this scenario, User 1 is closer to the base station than User 2. Therefore, h_1 causes less loss and User 1 has a higher total possible data rate. The base station allocates more power to User 2 based on the parameter α. User 1 can more easily remove the signal for User 2. Therefore, the total channel capacity for User 1 is based solely on the AWGN channel h_1 and the parameter α. User 2 does not perform SIC in this scenario. Therefore, the signal for User 1 counts as noise in the calculation of the channel capacity for User 2. Note that this scenario assumes that User 1 will have the ability to cancel the signal from User 2 completely.

$$R_1 = C\left(\alpha P |h_1|^2\right) \tag{10.5}$$

$$R_2 = C\left(\frac{(1 - \alpha)P |h_2|^2}{1 + \alpha P |h_2|^2}\right) \tag{10.6}$$

As can be seen in Figure 10.3, User 1 has a higher achievable rate due to a higher channel capacity and that the higher channel capacity is due to a lower loss in the channel for User 1. If the two users employ an OMA scheme where the two users must take turns having exclusive access to time frequency resources, the achievable rates are shown in the dashed line.

Figure 10.4   NOMA scenario for channel capacity analysis.

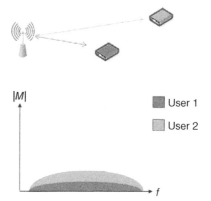

The achievable rates have a linear relationship because for every time-frequency resource that user 1 yields to user 2, user 2 gains a specific amount of throughput and user 1 loses a specific amount of throughput. However, if they transmit concurrently, then both users can enjoy a higher simultaneous data rate. For both the OMA and NOMA cases, if α is set to 1, then all resources are used for user 1, giving user 1 the maximum data rate for that channel and user 2 has a data rate of zero. Likewise, if α is set to 0, then all resources are used for user 2, giving user 2 the maximum data rate for that channel and user 1 has a data rate of zero.

10.2.2   Underlaying for Secondary Users

Underlaying was first mentioned in Chapter 5. Chapter 5 focused on *Opportunistic Spectrum Access*, and specifically precluded any discussion on underlaying as being out of scope. The IEEE 1900.1 standard provides a definition for **spectrum underlay** to include any case where a secondary user concurrently transmits on the same time-frequency resource as a primary user but keeps the resulting interference within established tolerable bounds [4]. The basic concept of underlaying is illustrated in Figure 10.5 from reference [5]. The secondary user transmits *under* the primary user in that the secondary user is transmitting at a lower relative power. The abbreviation "IT" stands for Interference Temperature. This is to say that the secondary transmission must be kept below a specific power to avoid interfering with the primary signal beyond that which the wireless link of the primary user can tolerate.

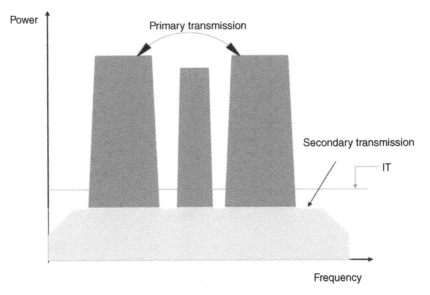

Figure 10.5   Underlaying for secondary users. *Source*: Tanab and Hamouda [5].

Underlaying for secondary users has not received much support from licensed wireless services or regulatory authorities. Reference [6] details the controversy surrounding secondary user access. Proprietors of licensed wireless services were not too keen on sharing their bandwidth in any capacity, especially when the secondary user is not even trying to avoid interference with primary transmissions.

10.2.3 Implementation Issues

The use of NOMA or underlaying requires that the receiver be able to cancel a portion of the received signal. That assumes that the receiver has a sufficiently accurate model for that portion of the signal. However, signal models as defined in the standards rarely take hardware impairments into account. Instead, the standards will impose limits on the allowable Bit Error Rate at some Minimum Detectable Signal as discussed in Chapter 3. The standards may also place a limit on the *Error Vector Magnitude* of the transmitter, which is a measure of how far from the ideal constellation points the transmission is. Hardware will always have imperfections. Such issues were explored in Chapter 3 in the Direct-Conversion Transceiver discussion. Reference [7] explores exactly that problem. What happens to the NOMA wireless link when real hardware impairments degrade the ability to cancel out signals?

Figure 10.6 from [7] illustrates the degradation in the Bit Error Rate Curve of a radio node, U2, required to cancel the signal for another node, U1.

Figure 10.6 Degradation of the BER curve for a receiver using SIC in the presence of amplifier non-linearity. *Source*: Selim et al. [7].

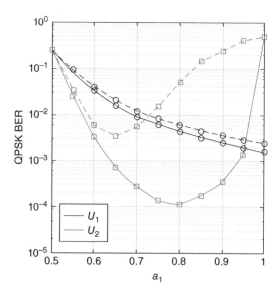

In this example, the amplifier at the transmitter has a non-linear transfer characteristic and the received signal also suffers from IQ-imbalance (the latter was discussed in Chapter 3). The independent axis is the power splitting parameter, "a". The SNR from AWGN for both nodes, not counting Multi-User Interference, is 25 db.

In Figure 10.6, the impairments degrade the BER of both nodes. Both nodes experience interference from the other. The node U2 attempts to cancel that interference. The node U1 does not. As the power splitting parameter goes to 1, the signal for node U2 is removed and U1 no longer experiences interference from U2. Figure 10.6 shows that a non-linear power amplifier adversely affects U2 far more than U1. This makes sense because U1 does not need a signal model to cancel U2. It is U2 that needs a signal model to cancel U1. When the signal model U2 is employing does not consider the non-linear power amplifier, the negative effects of the simple model are clearly demonstrated by Figure 10.6.

10.2.4 The Future of NOMA and Underlaying

As reference [6] states, the idea of transmitting one signal *under* another is not new. The concept of an interference channel is one where isolation between emitters is not complete, and different emitters interfere with one another in a shared spectrum resource. In reference [8], Carleial demonstrates that interference does not always reduce channel capacity. The interference channel has been modeled with one of the oldest analyses coming from Carleial in reference [9] in 1978. As was seen in Section 10.2.1, under certain conditions, interference can help two users improve throughput. Whether those conditions are realistic was explored in Section 10.2.3. There are difficulties achieving the cancelation required for optimal signal separation.

NOMA schemes have been proposed for 5G and the idea was discussed in literature. A small sampling of this literature shows the amount of research that went into the concept of using NOMA for 5G. Reference [10] from 2015, entitled *"Non-orthogonal multiple access for 5G: solutions, challenges, opportunities, and future research trends"*, has seen over 1100 citations. Reference [11] from 2017, entitled *"Application of Non-Orthogonal Multiple Access in LTE and 5G Networks"*, discusses using NOMA for 5G and has over 600 paper citations. These two are a small subset of the publications on the subject of using NOMA for 5G. Ultimately, NOMA schemes have not been adopted for 5G though NOMA remains an area of active research and may be a multiple access scheme that is adopted in the future.

10.3 Intelligent Collaborative Radio Networks

Significant research has been devoted to the concept of creating intelligent radios capable of operating anytime and anywhere utilizing the spectrum available to them at any given location. The notion that a single radio (or radio system) could accomplish this is very attractive for a number of reasons already discussed earlier in this book. First, individual radios acting independently to optimize spectral usage can make decisions decisively and require little or no cooperation with the environment outside of localized sensing. Additionally, if secondary users are acting completely independently there is the possibility that the radios could operate agnostic to the actual technologies used by the primary user. Deploying these independent radios, therefore, is a simple matter. They simply access spectrum that is not currently being used and no one is harmed during the process. The reality of this paradigm, however, is quite a bit harsher. To date, there are no existing cognitive radio systems that are able to independently operate within bands leased to traditional license holders. Whether this is a by-product of a missing technological breakthrough, market forces, or some combination of the two is irrelevant. The fact remains that the idea of "anytime anywhere" spectrum access outside of unlicensed bands is not, at the time of publishing, allowed by any major frequency coordination body.

In 2013 a DARPA program sought to improve on the state of these types of radios by creating the "Spectrum Challenge" [12]. The goal of this program was "to create protocols for software-defined radios that best use communication channels in the presence of other dynamic users and interfering signals" [12] and provided a $150 000 prize to the winning team. From the beginning, DARPA was clear that the goal was to develop new algorithms and methods for spectrum sharing and not simply building better physical layer techniques. Teams were able to write software defined radio code that ran on standardized set hardware in the ORBIT test facility located at Rutgers University shown in Figure 10.7 from reference [13]. These radios were co-located in a single lab providing a uniform and static radio channel between each node. The idea was that this combination of standardized hardware and static over-the-air channels ensured that all gains in spectrum efficiency would be the product of the competitors' algorithms and not a fluke of the physical radio mechanisms [14].

After the successful completion of the Spectrum Challenge, DARPA sought to build on the research by providing an additional means of optimization for the competitor's radios: collaboration. The goal of the new program, the Spectrum Collaboration Challenge (SC2), was to develop Collaborative Intelligent Radio Networks (CIRNs) that employed both reasoning *and* cooperation to ensure that all of the networks were able to carry their required data [15].

Though the use of the advanced Colosseum wireless emulation testbed discussed in Section 10.4.1, SC2 was able to put the radios though very specific and repeatable scenarios that tested the emergent behavior of the cooperating cognitive engines. Use of this emulated environment allowed for detailed analysis of algorithm behavior by repeatedly running identical deterministic scenarios in a dynamic emulated RF environment that allowed experiments to control the RF channel conditions, access policies, and data throughput targets very precisely.

Figure 10.7 400 identical radios in the ORBIT testbed at Rutgers. *Source*: Harbert [13].

In addition to this control, the DARPA team was able to clearly monitor and observe the behavior of the radios. For example, Figure 10.8 from reference [16] shows the spectrum allocations decided on by the radios during one particular run of an experiment simulating the data carried by different stores in a mall throughout the day [16]. During the morning (stage one) a coffee shop, shown in red, requires the most bandwidth during the early rush. As the day progresses, however, the bandwidth need for the other stores increases and networks intelligently collaborate to re-allocate the spectrum based on changing needs in stages two and three.

One of the key problems for competitors to solve in SC2 was balancing their own interests with that of the other teams. Scoring for each match was a combination of the quality of service provided by the team and was balanced by the quality of service provided by all teams [17]. In addition, teams had to avoid interfering with non-cooperative incumbents such as 802.11 networks and deal with malicious actors attempting to jam the channel. This complicated scoring rubric ensured that there was no easy answer that could be quickly arrived at through "expert feature building" [18]. The different dimensions to the scoring were meant to ensure that the radios could handle complex real-world scenarios. For example, it is unclear if mobile telephony operators would be willing to collaborate as freely as the teams in the competition might have if the only metric was the total data passed. Concerns over customer privacy, unpredictable quality of service for critical industries, and strategic business decisions could all put a damper on what and how much is shared in practice [19].

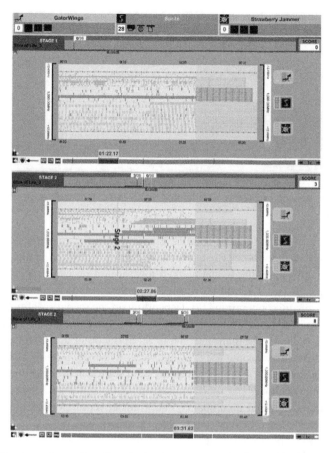

Figure 10.8 A spectrum allocation display for one scenario. *Source*: Tilghman [16].

Over the course of the competition, these dynamics emerged as teams were designing in special "competition modes" for their radios that sought to give them an edge [20]. Figure 10.9 from reference [16] shows an example of a scenario in which teams started by operating in a traditional frequency band plan and transition to a "full cooperation" mode halfway through. During this initial switchover, there is ample opportunity for teams to exploit the chaos during realignment. By analyzing the spectrum and current spectrum policies, some teams were able to adjust their behavior in ways that denied their closest competition the ability to score points thereby ensuring a win. There is no doubt that in a cutthroat industry like mobile telephony companies would employ similar tactics to give themselves an edge while still complying with the policies and regulations set forth by spectrum regulatory bodies.

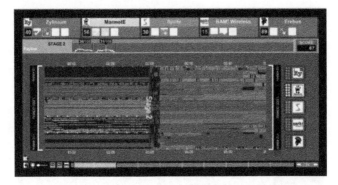

Figure 10.9 Networks switch from traditional channels to "Full Cooperation mode". *Source*: Tilghman [16].

Compared with the original DARPA Spectrum Challenge, the key difference with SC2 was the concept of collaboration. During execution of the challenge, DARPA scientists and the teams worked together on an evolving collaboration protocol that the teams would use during competitive matches to exchange information and ensure cooperation [17]. Depending on the scenario definition, different radios on each team were granted access to a low bandwidth "collaboration network" that connected them to some or all of the other teams in the scenario. Radios on this network could use the *Collaborative Intelligent Radio Network Interaction Language* (CIL), documented as an open source network protocol [21], to exchange information about spectrum sensing data, anticipated network moves, radio locations, detected incumbents, data transfer requirements, radio capabilities and more. Usage of this coordination network was not mandatory, but it would be impossible to score highly if the teams did not cooperate [17].

Overall, DARPA researchers considered the SC2 grand challenge to be a phenomenal success. They estimated that during the final matches, when the cognitive engines were operating at their peak, data was transmitted at rates two to three times what could be expected with traditional frequency band allocations [20]. While the research teams that competed in the program clearly recognize that there is a long way to go toward commercial adoption of these technologies, SC2 has demonstrated that collaboration between ostensibly "competing" interests can provide massive gains in performance for everyone.

10.4   Validation and Verification of Intelligent Radios

As the complexity and sophistication of spectrum sharing techniques grow, the interactions between heterogeneous decision-making engines will continue to grow at a rate that will quickly outstrip traditional methods of functional verification.

With cognitive engines making spectrum maneuvers based on a variety of constantly changing inputs, including the dynamic behavior of other cognitive processes, the explosion of potential configuration spaces quickly becomes intractable with respect to traditional wireless protocol analysis techniques. As discussed in detail in Chapter 2, the majority of radio devices used around the world must undergo specific testing procedures to ensure that they are operating within the bounds of the license they were developed for. Typically, this means nothing more than ensuring that the radio obeys the spectrum allocation masks mandated by the particular license. However, as both radio systems and licensing schemes become more complex additional testing will be required to ensure that devices can be certified. For example, testing the response of Dynamic Frequency Selection enabled 802.11 radios requires that a reference signal be fed to the receiver and then observe that the correct action is taken within a set time. Even though this is a relatively simple example, testing still requires several highly specialized pieces of equipment and requires examining 30 different tests against six different types of input waveforms to verify the correct behavior. As radios are allowed to take increasingly complex and non-deterministic cognitive actions, the number of test cases and scenarios will explode exponentially. This growth will require that new models for validation and verification of radio functionality be adopted and will be required.

Analysis and acceptance testing of cognitive engines and their radio waveforms is difficult for a number of reasons. The first major issue is ensuring that the cognitive engine adheres to the required policies in place at any given time. Given the fact that future dynamic spectrum access policies are expected to be capable of rapid changes in priority and allocations a cognitive engine must be able to demonstrate that the network parameters under its control are always in compliance with the policies as they change. This includes both the case of verifying that the cognitive engine made changes to comply with the policy and that it made those changes within a sufficient amount of time after the policy changed. The second major issue in acceptance testing is ensuring that the network parameters are actually changed when the cognitive engine sends updates in response to policy changes. If the network controller changes some physical layer parameters but the nodes in the network either miss the command or are unable to comply, the network will still be in violation of the policy, potentially interfering with incumbents. Acceptance testing of future cognitive radio devices and networks will therefore focus on two primary functions: the ability to determine a coherent parameter set given a policy set and the ability of the network to adhere to the defined parameter sets in a timely manner.

Verification of future cognitive radio systems adherence to policies and the ability of the network to react is not a trivial task. First, it can be assumed that the behavior of any given cognitive network would appear non-deterministic.

This is because there may be many overlapping policy restrictions in place and the goals of the cognitive engine may be dynamically changing based on the current network performance characteristics. Together, these two properties make the predictability of cognitive engine decisions very difficult without knowledge of the internal state of the controller. This means that traditional laboratory and field-testing exercises will likely not generate sufficient confidence that the network will operate in compliance with all policies. The chaotic nature of competing cognitive engines working to achieve independent goals weakens the value of these laboratory tests. Even though a single cognitive network may correctly comply with the policies during testing, interactions with other networks performing their own spectrum optimization algorithms will lead to edge cases that potentially force the network into a non-compliant state.

Due to the difficulties associated with ensuring adherence to spectrum sharing policies, it is clear that a new paradigm for vetting cognitive engine compliance is in order. While it is obvious that there will still be a need for laboratory and field-testing, it is clear that a middle step is required to provide assurance that a given cognitive engine both makes the correct decision given a particular environment and that the decision is made for the right reason. With the recent paradigm shift toward Software Defined Radios (SDRs), this middle ground can be achieved in a very effective manner by isolating the SDR code in a simulated or emulated environment and progressively increasing the fidelity of the environment. For example, the cognitive engine could be run as a stand-alone module during the first phase of testing by providing it with a set of policies and specific measurement data that change over time and observing its decisions to ensure compliance. The next stage could then introduce simulated RF links between the radio nodes running the deployed SDR code. Increasing the level of fidelity, the entire network code could then be moved to an emulated "hardware in the loop" testbed that presents the network with a number of sophisticated scenarios in an automated fashion. By emulating a real RF environment, the chaotic nature of the changing policies and measurements can be tested in a deterministic way. The radio network can be presented with the same set of stimuli several times, or individual stimuli can be perturbed over time to observe the resultant decisions of the engine. The primary benefit of the emulation environment is that several different types of radio networks, both DSA and incumbents, can be tested simultaneously in a deterministic and repeatable fashion. With a compliant implementation, reference designs for different approved radio networks can be saved and tested for compatibility against previously tested designs in a fully automated fashion, ensuring that there is no emergent behavior between different implementations that could accidentally cause policy infringements.

10.4.1 Case Study: The DARPA Colosseum

The DARPA Spectrum Collaboration Challenge (SC2) was a DARPA Grand Challenge that began in 2016 with the goal of producing breakthrough technologies to enable "a new spectrum paradigm that can help usher in an era of spectrum abundance" [5, 15]. Over the course of the three-year competition, 19 teams competed to build the best dynamic spectrum radios possible. The focus of the challenge was on collaboration between the radio networks themselves. Though the teams were in a competition with one another, the scoring algorithm was specifically designed to ensure that the teams collaborated to achieve the best possible utilization of the spectrum. When initially considering how to hold such a competition, DARPA scientists needed a way to ensure that different experiments and game matches were both fair and provided positive indication that a particular network had behaved correctly. The non-deterministic nature of the network interaction would make it very difficult to differentiate between radio systems that scored highly because they did the right thing as opposed to just getting lucky with respect to the operating environment. To solve this problem, DARPA turned to the Johns Hopkins University Applied Physics Lab (APL) in Laurel, MD to design a testbed that would allow for deterministic and repeatable experiments in a dynamic and chaotic RF environment consisting of both incumbent radios and experimental dynamic spectrum access radios.

To solve the problem of repeatedly scoring competitor radios in a fair and consistent manner, DARPA and the APL worked together to design the "Colosseum" which opened to competitors in April of 2017 [22]. The Colosseum is a testbed for intelligent radios that allows for the repeated and deterministic evaluation of performance in an endless number of network configurations and RF environments. The testbed is equipped with 128 "Standard Radio Nodes" (SRNs) which consist of a general-purpose processor, a GPU, and a 2×2 MIMO software defined radio. These nodes are connected to the heart of the system, a massive 256×256 channel emulator capable of fully re-combining any of the transmitted signals at any of the receivers according to a dynamic channel model developed for each experiment. Figure 10.10 from reference [23] shows a system-level view of the entire testbed. Competitors from around the globe were able to remote access the system via the internet, logging in to the "competitor access network", where they could access the SRNs that were assigned to them at any given time. Once they had developed the radios that were capable of running on these SRNs, competitors are able to queue up experiments that utilize the rest of the system components.

One of the key requirements of the Colosseum was that it could be "subdivided" to run multiple experiments simultaneously if the full 256×256 configuration was not needed. This allowed competitors to run small-scale experiments simultaneously and drastically improved the number of users that could simultaneously operate on the system.

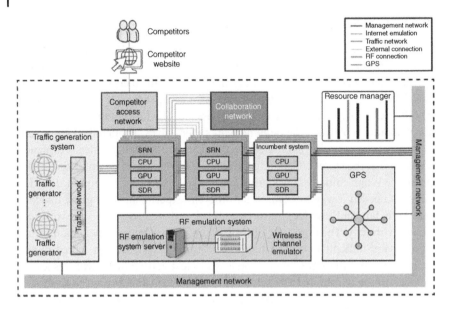

Figure 10.10 System-level view of testbed. *Source*: Coleman et al. [23].

However, this introduced significant logistical issues when it came to assigning resources in the system and required that a complex distributed system be built to satisfy all concurrent experiments running at any given time. For each experiment that is run, a scenario had to be loaded that defined the position of each radio node in the emulated environment over time, the types and amount of traffic the radios would be asked to pass, and a model for the full NxN RF channel matrix between each node, and the parameters defining the incumbent radio behavior. These values would vary over the course of each experiment and so everything had to be allocated and tracked in real time to determine the current state of any given experiment and determine when resources were going to be needed to satisfy a given test scenario. While this additional effort proved to be a non-trivial task [24], it serves as an important model for future testing and emulation environments. It is clear that testbeds such as the colosseum will become a bottleneck for evaluating radio behavior if they cannot be subdivided in a way that provides the ability to effectively subdivide the emulation resources.

When evaluating cognitive systems and their ability to find available spectrum in traditionally allocated bands the primary concern is potential interference generated against the incumbent users in the band. For this reason, the Colosseum has the ability to include incumbent users in the simulated and emulated scenarios and incorporates an advanced interference detection model that provides feedback on which specific radio was interfering with the incumbent at any time. This allows testing agencies to provide specific feedback to radio and algorithm designers about the RF landscape leading up to the interference event thereby providing the necessary information to improve the algorithm. Figure 10.11 from reference [25] shows an example of the algorithm used by Colosseum to indicate whether an incumbent is interfered with at any given time. Once the interfering power crosses the set threshold, teams are penalized until that interference drops below the required value for a set amount of hysteresis.

For the Spectrum Collaboration Challenge, a number of incumbent radios were designed to utilize the same Standard Radio Node that the competitors ran their own modems with [25]. For the sake of the training and development phases leading up to the competition, sharing a common platform with the radios under test drastically simplified the resource contention problem by pulling incumbent radios from the common pool of SRNs. This model provided the maximum flexibility, allowing the full channel emulator to be assigned to SRN radios at any given time. However, because the channel emulator at the heart of the Colosseum operates across a wide frequency band and in real time [26], it is possible to connect true incumbent hardware into the emulation environment to determine the interaction behavior of intelligent systems against the actual incumbents they will encounter in the field. This tradeoff is a critical parameter when designing future solutions for validation and testing new spectrum sharing radios, as the size of the

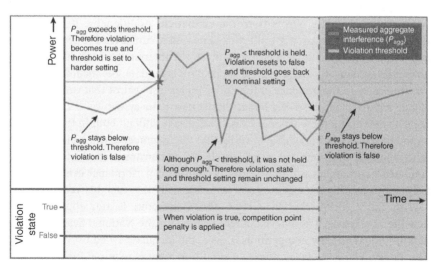

Figure 10.11 An example of the incumbent interference detection algorithm used by Colosseum. *Source*: Yim et al. [25].

network that can be tested will be limited based on the number of incumbent radios that need to be installed.

In order to evaluate the dynamic spectrum radios effectively, a complex-data generation system was developed to provide representative traffic to the user radios [26]. This system was capable of generating IP packet data that represented a number of potential traffic scenarios such as streaming voice and video, web browsing, and real time tactical updates. In addition, the radios at the receiving end of the traffic would report the packets that they received back to the traffic generation system for evaluation. Figure 10.12 from reference [27] shows an example of one traffic generator flow through the system. Data travels from the generator to the SRN, is passed over the RF link by the competitors' modem and returned to the traffic sink where it is checked by the evaluator.

With this closed loop scoring system, the traffic generator could collect statistics about the information carrying capacity of the link between each pair of radios in the network including things such as packet error rate, overall throughput, and average latency. Because the traffic sinks and sources could be any service running in a container, this system allowed for the measurement of actual system level effects of the radio's performance. By utilizing real networking stacks, the effects of PDU retransmissions, failed acknowledgements, buffer overflows and other difficult to simulate phenomena are all accurately measured at the system level. Compared with traditional traffic evaluation tools that simply check whether a PDU was received or not, this system can measure video and audio quality of service, web browsing latency, and more with the finest level of granularity while leaving competitors free to optimize the data transfer in any way they saw fit without worrying about secondary effects from the emulation. Additionally, if incumbent radios are designed to carry IP traffic this same system can be used to generate both quantitative and qualitative measurements of the interference introduced by the new radio systems. By analyzing the application layer data, the incumbents were asked to carry in the presence of the radios under test against that run in a standalone environment the direct impact of the new radio system can be measured. Directly measuring these types of causal statistics is difficult both in the field and on the bench as it is impossible to verify that the incumbents will be in the same state when starting each test run. This clearly demonstrates the benefits of the emulated environment as it allows directly comparing performance in a repeatable fashion by ensuring the experimental state is exactly the same between runs.

With the DARPA challenge over, the Colosseum is transitioning to a new life as a research and development testbed for the National Science Foundation [28]. While the competition is over, the lessons learned over the three-year project can provide significant insight into what is required for testing the behavior of future intelligent radio systems.

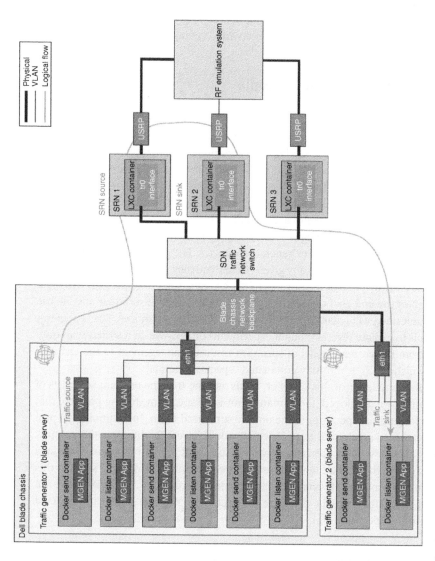

Figure 10.12 The data flow of traffic through the Colosseum. *Source:* Curtis et al. [27].

It is very likely that a system similar to the Colosseum will one day be used by regulatory bodies to certify intelligent radios by presenting them with a number of predefined scenarios in an automated fashion. These scenarios can include incumbents as well as other radio networks that will be operating in the same spectrum spaces to ensure that there is full policy compliance when a complex RF environment exists.

10.5 Spectrum Sharing Utopia

All research in cognitive radio and dynamic spectrum access is distributed across different levels of expected incumbent compatibility. When inventing and researching optimal spectrum usage, strategies to accomplish optimality range everywhere between "blue sky" efforts that assume all spectrum users are capable of coordination and agility to "worst case" strategies that assume only the current user is agile and must operate with perfect adherence to interference policies. Making different assumptions about the makeup and functionality of heterogeneous networks can lead to wildly different strategies and optimization techniques. However, the goal of this research is not simply to generate strategies. The goal is to ensure that spectrum is optimally allocated in any given "time frequency position" resource block at any time and any place in the universe! This, though, is not truly the goal. This simply enables the end goal of spectrum sharing which is to ensure that anyone can communicate at will regardless of the available spectrum resources. The goal of this section is to explore several thought experiments on how we, as a society, can transition from the current model of spectrum resource management to the end goal state, which is a free and fully automated "spectrum utopia".

For the remainder of this Chapter, let us assume that the brilliant scientists of the world have developed, for humanity, an ideal spectrum sharing protocol that allows for optimal spectrum usage on any supported radio. In addition, this protocol is granted royalty free to anyone that would have it for the good of humanity. This could allow for a scenario in which spectrum regulators across the globe decide, overnight, that all new radios and spectrum allocations must support this new "utopia protocol". Even in this fantastic imagination, we as a society have not arrived at spectrum utopia: there are still factors in play that hinder unrestricted access to the spectrum.

10.5.1 Major Hurdles for Spectrum Utopia

In order to achieve this spectrum utopia, several large hurdles must be overcome. While there are certainly technical challenges, which are the primary focus of the other chapters in this book, there are also political and economic challenges that cannot be solved by "inventing a better method". When we free ourselves from looking at the technical side of the spectrum regulations, it is clear that these primary hurdles to a spectrum utopia that cannot be solved away with "simple" technical fixes are largely the political and economic forces surrounding spectrum regulation: incumbent hardware, current license holders, and the limited number of bands for global harmonization.

The first primary hurdle, fielded incumbent hardware, may seem at first glance like a technical hurdle and not necessarily an economic or political one. However, given the fact that any incumbent theoretically could be replaced, even if the costs were astronomical, there is an economic and political component to this problem. The issue with incumbent hardware is clear: because these currently fielded radios do not support the utopia protocol they cannot share their spectrum with new radios. It is important to note that most, if not all, spectrum licensing schemes provide users with a right to be "free from interference". While, in theory, this is not the same policy as "exclusive rights to transmit energy on a channel" the two policies have become synonymous as we lack the technical means to enforce the first policy through any method other than the second. The "state of practice", then, is that the vast majority of fielded radio media access control (MAC) mechanisms are dependent on the idea that their radio system is the one allowed to transmit on the channel. Therefore, regardless of the coordination mechanisms that might be used to build interleaved and overlay networks the existing radios will always have the potential to be interfered with because their media access control (MAC) mechanisms will likely be hampered by any additional energy in the band. The effect of these incumbent radios is clear: until they are replaced or upgraded the spectrum, they inhabit will be off limits to spectrum utopia. Either this replacement and upgrade process will occur naturally, which may take generations, or it will be prohibitively expensive and, as such, require a major force of will – either political or altruistic – to compress the timeline and usher in an era of spectrum utopia.

The second primary hurdle to the spectrum utopia is the current licenses that grant exclusive use of the spectrum to non-governmental entities, often at great cost. Because private entities now effectively "own" large swaths of spectrum, there are very few governing bodies that "take back" that spectrum without a major change in their system of property rights. Even if the "utopia protocol" were free to everyone and incumbent radios were outlawed, these licenses would still provide the holders with exclusive rights to that spectrum and, because they paid for it, will require the ability to extract value from that spectrum. This hurdle is likely the easiest to overcome in some respects, while clearly it is the hardest in others. On the one hand, it would be simple for governments to begin the process of "buying back" leases over time and allocating it to the utopia protocol, but market forces would quickly drive up the cost of those buybacks. If a corporation knows that anyone, especially a government, is keen on buying their assets you can bet the asking price will be very high. While there are certainly steps that regulatory bodies could take to make these spectrum rights less valuable, it is not clear that they would even want to. Spectrum auctions have proved to be a very lucrative method for raising public funds, though it could be argued that this value would also drastically decrease once the "utopia protocol" was available to everyone. While it's impossible to predict the market forces that will drive the value of spectrum licenses in a world where the utopia protocol exists, it is clear that the owners of these rights will not willingly give them up without an effective "return on investment" for the property they paid for. Any path to spectrum utopia must consider how these license holders could be convinced to give up their right to exclusivity.

The final major hurdle in our spectrum utopia is the concept of global band harmonization. If we consider what happens starting on "day zero" of the utopia protocol being released, there will have to be some path that takes from our current global radio regulations toward the ideal state of there being no need for radio regulations at all. What is most clear is that there will need to be a gradual adoption of the utopia protocol. Undoubtedly, the ISM and other unlicensed bands will be the first to see major adoption and, as legacy spectrum becomes available, it could be devoted to exclusive use by radios that implement the protocol. However, this transition will at best provide a piece-meal solution from a geographic perspective until the whole world has declared a particular channel "utopified". In order for a critical mass of radios to be developed, there must be sufficient spectrum available to support the protocol, but in order for governments to commit those spectrum resources there must be a critical mass – a classic deadlock. The limited number of bands available for global harmonization and the lack of clear benefit from "freeing up" unlicensed spectrum will dramatically slow the pace of adoption for spectrum utopia. For example, if a particular technology loses utility on a global scale it can be phased out such as happened with analog television. However, because the bands were not harmonized in the first place, freeing up those bands leaves large gaps in the spectrum that do not overlap globally. Regulators have a choice: leave these gaps unallocated and wait for future phase outs that enable overlap or re-allocate the empty spectrum to new technologies which are now "incumbent". Given the ever-increasing demand for spectrum, history has shown that leaving spectrum bands unused for the prospect of future harmonization has not proven popular. This means that if the goal is true global harmonization of bands, the pace of progress toward that goal will be glacially slow, relying on the random effects of technological obsolescence in currently un-coordinated bands.

10.5.2 Pathways to an Optimally Utilized Future

For the remainder of this chapter, we will focus on three potential paths could conceivably take society from the release of the utopia protocol to our imagined future where the entire spectrum is "unlicensed" and all wireless applications exist harmoniously. These three development paths each focus on a single primary stakeholder pushing for change that represent the different interests in spectrum usage. The stakeholders in each of the three scenarios are the public sector, the industrial sector, and the private sector. In each case, there is a different set of people whose actions provide a plausible path to spectrum utopia.

In the first scenario, governments and regulatory bodies will provide the primary impetus to spectrum utopia. The timeline from the initial release of the utopia protocol to a fully unlicensed world takes shape slowly but grows quickly as the effects of adoption and harmonization translate to improved economic success for participating governments and the laggards follow suit. Initially, regulatory agencies could begin by opening up all currently unlicensed allocations to radios that implement utopia protocol. Now enabled, a critical mass of use cases and radio networks would demonstrate the clear benefits of optimally allocated spectrum resources allowing regulatory bodies to designate future spectrum allocations exclusively to utopia protocol. From here, potential issues with this path become more apparent. How is it that hardware incumbents and exclusive licensees can be made to participate? Without the political appetite for retroactively amending property negotiations, incumbents which currently hold the vast majority of allocations will be free to continue under-using their spectrum until it makes economic sense otherwise. It is at this point that governments must decide if they wish to use regulatory muscle to "claw back" the spectrum from rights holders. While there have been precedents for removing physical property rights when misused by the owners in the past, it is clear that this "eminent domain" style of spectrum recovery will be extremely unpopular with lease holders. In this scenario, the hurdle of incumbent hardware is overcome through regulatory sunsets on allowed spectrum usage – something which may prove unpopular in the commercial sector. Likewise, the hurdle of existing spectrum licenses will either be revoked outright or made economically worthless as cheaper services on the unlicensed bands replace legacy technologies. Finally, the timeline for the global harmonization hurdle can be shortened because the regulators are the impetus behind the change. The major drawback of this plan is that regulators must make a difficult choice: delay the onset of spectrum utopia by allowing inefficient operators and leaseholders to continue "business as usual" or amend contractual licenses for the greater good. The benefits of this model, though, are that access to the spectrum is ensured as a regulatory function of the government and cannot be easily taken away. Spectrum access becomes a "human right" in the sense that everyone is able to use what they need by force of law, rather than through an economic exchange.

The second potential path forward involves the minimum amount of political will, as it relies entirely on the economic forces of the industrial sector to drive society toward spectrum utopia. In this scenario, we can imagine that slight modifications to the utopia protocol are implemented that allow for a "fee" to be attached to each allocation.

Now as each radio accesses a certain portion of the spectrum, they will be charged according to the fee set by the license holder. While nothing is precluding this type of model from being used today, the introduction of the utopia protocol allows this to happen in real time, allocating and tracking usage all while adding to the balance owed. This style of sharing would likely not be effective in a consumer facing subscription model due to the unpredictable costs of access. Instead, it is more likely that license holders would arrange "peering agreements" similar to those already in place between internet backbone carriers. In the case of peering agreements, pairs of resource holders agree to pay a fixed rate to use one another's resource. The actual billing of these charges happens periodically, but in practice optimal allocation would mean that the both parties used the same amount of the resource and therefore their charges are balanced out – no one owes anything! In this case, the hurdles are easily overcome due to the massive economic benefits to lease holders. If it means the ability to carry more customer data, field upgrades of legacy equipment will be prioritized eliminating the incumbent radio problem. Likewise, the license holders are no longer a problem as they are the primary drivers of the plan: they are earning economic benefit from their property and their leases remain intact. Only the final hurdle, global harmonization, remains. This hurdle is also overcome by our new "instant marketplace for spectrum" because the primary barrier to global harmonization is coordinated phase outs of incumbent technologies, the "gold rush" of peering agreements will provide a strong economic benefit for incumbent equipment operators to phase out their old technologies in favor of utopia protocol. While this path toward spectrum utopia has the minimal amount of political and economic disruption, it also completely locks out the idea of "spectrum commons". If the spectrum is a natural resource, this model devotes the entirety of that resource toward private parties and does not provide in any way for the "greater good". In addition, as more spectrum is allocated toward leased and economic uses during the harmonization phase, governmental and military access to the spectrum will, by definition, shrink. This limits the government's ability to allocate its own spectrum for the utopia protocol as a balance against the property rights granted to industry.

The final path to utopia we will discuss here is focused on a scenario in which the private sector, or citizenry, work together for the greater good to bring about spectrum utopia. This scenario follows similar to the industry scenario, in which a de facto monopoly on spectrum access is built by a private entity. In this case, however, the primary function of that entity would be to ensure free access for the greater good. Here we envision a model in which a nonprofit entity is founded with the goal of buying up as many spectrum leases as possible and declaring them for use of the utopia protocol. During the initial phases, this foundation would require considerable resources in order to establish a foothold large enough to reach a critical mass of deployed hardware.

It is likely that this initial capital will need to come from a hyper-wealthy philanthropist with a personal cause. In much the same way that Andrew Carnegie devoted a large portion of his personal wealth to the cause of public libraries, a future titan of industry could spur similar advancements in spectrum access which would be later sustained through a combination of public funds and private endowments. Similar to the industrial sector scenario, the private sector scenario avoids the hurdle of license holders by ensuring that they are economically compensated for their property. Like the public sector scenario, the private sector scenario carries the voice of the people and ensures the political will necessary to make global band harmonization a priority. The weakest hurdle for this scenario is the problem of incumbent hardware. It is unlikely that a privately funded entity with no profit motive could accumulate the necessary capital to buy out incumbent hardware at the scale necessary for total spectrum utopia. Overall, this scenario provides for a middle path between the first two: there are no major changes to property rights or existing spectrum leases, but free access to the spectrum is still guaranteed for the people. The major drawback is that it relies on the existence of a group of individuals who possess both the force of will and financial resources necessary to build a lasting and durable trust capable of maintaining the notion of free access.

By comparison each of the three paths to spectrum utopia covered here have their pros and cons related to three major factors: provided rate of progress, required ideological shifts in government, and resulting open access levels. Table 10.1 shows a side by side comparison of the three paths including the pros and cons of each.

In reality the more likely scenario is that if we do achieve spectrum utopia, it will be through a combination of these paths. None of the milestones toward achieving utopia in any of the presented scenarios are mutually exclusive. For example, there is nothing stopping the industrial and private sector paths from peacefully coexisting resulting in a mixed set of outcomes that provide benefits to both the "greater good" and the industrial economic base.

10.6 Conclusion

This book is a result of the authors asking a simple question: "why is wireless coexistence hard?". Though this question is simple to ask, almost certainly while clustered around the bar at the barbeque joint across the street from the office, the answer is horrifyingly complex. The interwoven set of issues that arise from the mix of political, commercial, and technical interests present a knot whose complexity approaches levels that can only be described as "Gordian".

Table 10.1 Table TBD.

	Rate of progress	Benefit to common good	Required ideology shifts	

Public	Medium	High	High	• Spectrum access is free and open
				• Potential political fallout
				  – Loss of auction revenue
				  – Eminent domain issues
				• Little incumbent phaseout incentive

Industrial	High	None	Low	• No major regulation changes • Property "value" is left in tact • No benefit to common good
Private	Low	High	Medium	• No major regulation changes • Requires motivated individuals • Slow to phase out incumbents

This book was designed to lead the reader on a journey that would extend beyond the "how" of different coexistence standards into the "why" and, perhaps more importantly, the "why not".

A competent engineer, provided access to the specifications, could undoubtedly create a radio that complied with all of the rules and was able to operate seamlessly with other compliant radios. It is unlikely that in this process, however, the engineer would understand the *why* of those rules. Why does this spectral mask look like the skyline of Cleveland, Ohio? Why am I only allowed to transmit for 147 ms at a time? Why not 148 ms? As we delved deeper into these types of questions it quickly became clear that the answers were never straight forward. Perhaps you must limit your power levels in certain bands to prevent interference with submarines, even if you are in the desert. It may be the case that you cannot use channel 8 on your radio dial because it would interfere with a piece of equipment critical to the national infrastructure that was designed and built before highly selective filters were available. It is also entirely possible that a peace treaty signed to end a long-forgotten war prevents you from legally using a particular frequency to carry voice traffic. Whatever the reasons for seemingly odd protocol decisions, it is clear that wireless coexistence is a messy topic. By understanding the different limitations placed on a radio operator by political forces, historical engineering practices, and the current state-of-the-art in hardware an engineer can develop a radio that is able to comply to the spirit of the standards and not just to the letter. In this way, through experimentation and exploration in the unlicensed bands, new and better methods for coexistence can be constantly iterated on providing more efficient use of a finite resource.

Given this myriad of potential influences on any given standard, we wrote this book to endow the reader with the ability to evaluate standards though the different lenses that directly impact when and how any radio system is allowed to access the electromagnetic spectrum. After starting with a brief introduction in wireless theory to ground the reader's perspective, a history of technological progress and regulations governing access to electromagnetic transmissions provides insight into how we got to where we are today. By then building up the core technical knowledge surrounding how radios can access and sense that spectrum, the reader gains an understanding of the scope and limitations on the techniques that may be employed to communicate. Finally, by presenting the state-of-the-art in intelligent cooperative radio techniques the reader can get a feel for the "art of the possible" when it comes to maximizing use of the spectrum given the constraints we face from a political and historical point of view. Armed with this knowledge, we could then proceed with coverage of the existing standards that are just starting to scratch the surface of what is possible in the field of spectrum sharing and opportunistic spectrum allocation. Undoubtedly the success of these standards will provide a strong basis for future regulations that dedicate precious spectrum resources solely to intelligent radios that are capable of maximizing their benefit.

It is our hope that by maintaining a holistic view of the problem, the reader can objectively evaluate current and future coexistence strategies for practical applications in a complex regulatory landscape.

References

1 Voicu, A.M., Simić, L., and Petrova, M. (2018). Survey of spectrum sharing for inter-technology coexistence. *IEEE Communications Surveys & Tutorials* 21 (2): 1112–1144.

2 Xiong, X., Xiang, W., Zheng, K. et al. (2015). An open source SDR-based NOMA system for 5G networks. *IEEE Wireless Communications* 22 (6): 24–32.

3 Vaezi, M., Schober, R.D.Z., and Poor, H.V. (2019). Non-orthogonal multiple access: common myths and critical questions. *IEEE Wireless Communications* 26 (5): 174–180.

4 IEEE. (2019). IEEE Standard for Definitions and Concepts for Dynamic Spectrum Access: Terminology Relating to Emerging Wireless Networks, System Functionality, and Spectrum Management, IEEE Std 1900.1-2019.

5 Tanab, M.E. and Hamouda, W. (2017). Resource allocation for underlay cognitive radio networks: a survey. *IEEE Communications Surveys & Tutorials* 19 (2): 1249–1276.

6 Marcus, M.J. (2007). The underlay/overlay controversy. *IEEE Wireless Communications* 14 (5): 4–5.

7 Selim, B., Muhaidat, S., Sofotasios, P.C. et al. (2019). Radio-frequency front-end impairments: performance degradation in nonorthogonal multiple access communication systems. *IEEE Vehicular Technology Magazine* 14 (1): 89–97.

8 Carleial, A. (1975). A case where interference does not reduce capacity (Corresp.). *IEEE Transactions on Information Theory* 21 (5): 569–570.

9 Carleial, A. (1978). Interference channels. *IEEE Transactions on Information Theory* 24 (1): 60–70.

10 Dai, L., Wang, B., Yuan, Y. et al. (2015). Non-orthogonal multiple access for 5G: solutions, challenges, opportunities, and future research trends. *IEEE Communications Magazine* 53 (9): 74–81.

11 Ding, Z., Liu, Y., Choi, J. et al. (2017). Application of non-orthogonal multiple access in LTE and 5G networks. *IEEE Communications Magazine* 55 (2): 185–191.

12 DARPA. (2020). DARPA Spectrum Challenge. https://www.darpa.mil/program/ spectrum-challenge (accessed 26 May 2020).

13 Harbert, T. (2014). Radio Wrestlers Fight It Out at the DARPA Spectrum Challenge. IEEE Spectrum News. https://spectrum.ieee.org/telecom/wireless/radio-wrestlers-fight-it-out-at-the-darpa-spectrum-challenge (accessed 26 May 2020).

14 DARPA. (2014). Spectrum Challenge Final Event Helps Pave The Way For More Robust, Resilient And Reliable Radio Communications. https://www.darpa.mil/ news-events/2014-04-02a (accessed 26 May 2020).

15 DARPA. (2020). What is the Spectrum Collaboration Challenge?, DARPA. https:// archive.darpa.mil/sc2/about (accessed 26 May 2020).

16 Tilghman, P. (2019). If DARPA Has Its Way, AI Will Rule the Wireless Spectrum. *IEEE Spectrum*, 28 May 2019.

17 DARPA. (2019). Spectrum Collaboration Challenge Rules. https://archive.darpa. mil/sc2/wp-content/uploads/2019/02/SC2-Rules-Document-V3.pdf (accessed 26 May 2020).

18 Koziol, M. (2019). Darpa Grand Challenge Finale Reveals Which AI-Managed Radio System Shares Spectrum Best. IEEE Spectrum Tech Talk. https://spectrum. ieee.org/tech-talk/telecom/wireless/aimanaged-radio-systems-duked-it-out-to-see-which-one-could-share-spectrum-the-best-in-the-spectrum-collaboration-challenge-finale (accessed 26 May 2020).

19 Koziol, M. (2019). AI-enabled spectrum technology: What's next? *IEEE Spectrum* 56 (12): 6–6.

20 Koziol, M. (2019). DARPA's Grand Challenge Is Over – What's Next for AI-Enabled Spectrum Sharing Technology?. IEEE Spectrum Tech Talk. https://spectrum.ieee. org/tech-talk/telecom/wireless/with-darpas-spectrum-collaboration-challenge-completed-whats-the-next-step-for-spectrum-sharing-technologies (accessed 26 May 2020).

21 DARPA. (2019). Spectrum Collaboration Challenge (SC2) CIRN (Collaborative Intelligent Radio Network) Interaction LanguageGit Repository. DARPA. https:// github.com/SpectrumCollaborationChallenge/CIL (accessed 26 May 2020).

22 DARPA. (2017). World's Most Powerful Emulator of Radio-Signal Traffic Opens for Business. DARPA. https://www.darpa.mil/news-events/2017-04-21 (accessed 7 June 2020).

23 Coleman, D.M., McKeever, K.R., Mohr, M.L. et al. (2019). Overview of the colosseum: the World's largest test bed for radio experiments. *Johns Hopkins APL Technical Digest* 35 (1): 4–11.

24 Mok, J.W., Hom, A.L., Uher, J.J., and Coleman, D.M. (2019). The resource manager for the Defense Advanced Research Projects Agency Spectrum collaboration challenge test bed. *Johns Hopkins APL Technical Digest* 35 (1): 34–41.

25 Yim, K.J., McKeever, K.R., and Barcklow, D.R. (2019). Incumbent radio systems in the Defense Advanced Research Projects Agency Spectrum collaboration challenge test bed. *Johns Hopkins APL Technical Digest* 35 (1): 49–57.

26 Barcklow, D.R., Bloch, L.E., Sweeney, S.W. et al. (2019). Radio frequency emulation system for the Defense Advanced Research Projects Agency Spectrum collaboration challenge. *Johns Hopkins APL Technical Digest* 35 (1): 69–78.

27 Curtis, P.D., Plummer, A.T.J., Annis, J.E., and La Cholter, W.J. (2019). Traffic generation system for the Defense Advanced Research Projects Agency Spectrum collaboration challenge. *Johns Hopkins APL Technical Digest* 35 (1): 58–68.

28 DARPA. (2019). World's Most Powerful RF Emulator to Become National Wireless Research Asset. DARPA. https://www.darpa.mil/news-events/2019-09-03 (accessed 6 June 2020).

Index

Wireless Coexistence: Standards, Challenges, and Intelligent Solutions, First Edition.
Daniel Chew, Andrew L. Adams, and Jason Uher.
© 2021 The Institute of Electrical and Electronics Engineers, Inc.
Published 2021 by John Wiley & Sons, Inc.

Printed and bound by CPI Group (UK) Ltd, Croydon, CR0 4YY

25/01/2023

03183928-0003